CMS/CAIMS Books in Mathematics

Volume 6

Series Editors

Karl Dilcher
Department of Mathematics and Statistics, Dalhousie University, Halifax, NS, Canada

Frithjof Lutscher
Department of Mathematics, University of Ottawa, Ottawa, ON, Canada

Nilima Nigam
Department of Mathematics, Simon Fraser University, Burnaby, BC, Canada

Keith Taylor
Department of Mathematics and Statistics, Dalhousie University, Halifax, NS, Canada

Associate Editors

Ben Adcock
Department of Mathematics, Simon Fraser University, Burnaby, BC, Canada

Martin Barlow
University of British Columbia, Vancouver, BC, Canada

Heinz H. Bauschke
University of British Columbia, Kelowna, BC, Canada

Matt Davison
Department of Statistical and Actuarial Science, Western University, London, ON, Canada

Leah Keshet
Department of Mathematics, University of British Columbia, Vancouver, BC, Canada

Niky Kamran
Department of Mathematics and Statistics, McGill University, Montreal, QC, Canada

Mikhail Kotchetov
Memorial University of Newfoundland, St. John's, Canada

Raymond J. Spiteri
Department of Computer Science, University of Saskatchewan, Saskatoon, SK, Canada

CMS/CAIMS Books in Mathematics is a collection of monographs and graduate-level textbooks published in cooperation jointly with the Canadian Mathematical Society-Societé mathématique du Canada and the Canadian Applied and Industrial Mathematics Society-Societé Canadienne de Mathématiques Appliquées et Industrielles. This series offers authors the joint advantage of publishing with two major mathematical societies and with a leading academic publishing company. The series is edited by Karl Dilcher, Frithjof Lutscher, Nilima Nigam, and Keith Taylor. The series publishes high-impact works across the breadth of mathematics and its applications. Books in this series will appeal to all mathematicians, students and established researchers. The series replaces the CMS Books in Mathematics series that successfully published over 45 volumes in 20 years.

CMS
SMC

CAIMS
SCMAI

Alexander Melnikov

A Course of Stochastic Analysis

 Springer

Alexander Melnikov
Department of Mathematical
and Statistical Sciences
University of Alberta
Edmonton, AB, Canada

ISSN 2730-650X   ISSN 2730-6518 (electronic)
CMS/CAIMS Books in Mathematics
ISBN 978-3-031-25325-6   ISBN 978-3-031-25326-3 (eBook)
https://doi.org/10.1007/978-3-031-25326-3

Mathematics Subject Classification: 60, 62

© The Editor(s) (if applicable) and The Author(s), under exclusive license to Springer Nature Switzerland AG 2023

This work is subject to copyright. All rights are solely and exclusively licensed by the Publisher, whether the whole or part of the material is concerned, specifically the rights of translation, reprinting, reuse of illustrations, recitation, broadcasting, reproduction on microfilms or in any other physical way, and transmission or information storage and retrieval, electronic adaptation, computer software, or by similar or dissimilar methodology now known or hereafter developed.

The use of general descriptive names, registered names, trademarks, service marks, etc. in this publication does not imply, even in the absence of a specific statement, that such names are exempt from the relevant protective laws and regulations and therefore free for general use.

The publisher, the authors, and the editors are safe to assume that the advice and information in this book are believed to be true and accurate at the date of publication. Neither the publisher nor the authors or the editors give a warranty, expressed or implied, with respect to the material contained herein or for any errors or omissions that may have been made. The publisher remains neutral with regard to jurisdictional claims in published maps and institutional affiliations.

This Springer imprint is published by the registered company Springer Nature Switzerland AG
The registered company address is: Gewerbestrasse 11, 6330 Cham, Switzerland

Preface

The foundation of the modern probability theory was done by A. N. Kolmogorov in his monograph appeared as "Grundbegriffe der Wahrscheinlichkeitsrechtung" in 1933. Since that time we use the notion of a probability space $(\mathbf{\Omega}, \mathcal{F}, \mathbf{P})$, where $\mathbf{\Omega}$ is an abstract set of elementary outcomes ω of a random experiment, $\mathcal{F}$ is a family of events, and P is a probability measure defined on $\mathcal{F}$. In that book, there is a special remark about the property of independence as a specific feature of the theory of probability. Due to this property, one can use deterministic numerical characteristics like mean and variance to describe the behavior of families of random variables. It was later recognized that dynamics of random events can be exhaustively determined considering t and ω together and a random process as a function of these two variables on the basis of information flow F_t, $t \geq 0$.

The fruitful idea was extremely important for a general theory of random processes initiated in the middle of the twentieth century. The central notion of this theory is a stochastic basis $(\mathbf{\Omega}, \mathcal{F}, (\mathbf{F}_t), \mathbf{P})$, i.e. a probability space equipped with an information flow or filtration (F_t). In such a setting, deterministic numerical characteristics induced by the independence property are replaced by their conditional versions with respect to filtration (F_t). So, a predictability has appeared as a driver of extension of stochastic calculus to the biggest possible class of processes called semimartingales. These processes admit the full description in predictable terms. Moreover, they unify processes with discrete and continuous time. As a result, we arrive to a nice transformation of the theory of probability and stochastic processes to a wider area which is called stochastic analysis.

The primary goal of the book is to deliver basic notions, facts, and methods of stochastic analysis using a unified methodology, sufficiently strong and complete, and giving interesting and valuable implementations in mathematical finance and statistics of random processes. There are a lot of examples considered to illustrate theoretical concepts discussed in line with problems for students aligned with material. Moreover, the list of supplementary problems with hints and solutions, covering both important theoretical statements and purely technical problems intended to motivate a deeper understanding of stochastic analysis is provided at the end of the book. The book can be considered as a textbook for both senior undergraduate and graduate courses on stochastic analysis/stochastic processes. It certainly can be helpful for undergraduate and graduate students, instructors as well as for experts in stochastic analysis and its applications.

The book is based on the lecturers given by the author in different times at Lomonosov Moscow State University, State University-Higher School of Economics, University of Copenhagen, and at the University of Alberta. The author is grateful to his Ph.D. students Ilia Vasilev, Andrey Pak, and Pounch Mohammadi Nejad at the Mathematical Finance Program of the University of Alberta for their kind help preparation of this book.

Edmonton, Canada   Alexander Melnikov

Contents

1 Probabilistic Foundations 1
    1.1   Classical theory and the Kolmogorov axiomatics 1
    1.2   Probabilistic distributions and the Kolmogorov consistency
           theorem 6

2 Random variables and their quantitative characteristics 13
    2.1   Distributions of random variables 13
    2.2   Expectations of random variables 18

**3 Expectations and convergence of sequences of random
variables** 21
    3.1   Limit behavior of sequences of random variables
           in terms of their expected values 21
    3.2   Probabilistic inequalities and interconnections between
           types of convergence of random variables... 26

4 Weak convergence of sequences of random variables 31
    4.1   Weak convergence and its description in terms
           of distributions 31
    4.2   Weak convergence and Central Limit Theorem 34

**5 Absolute continuity of probability measures and conditional
expectations** 41
    5.1   Absolute continuity of measures and the Radon-Nikodym
           theorem 41
    5.2   Conditional expectations and their properties 43

6 Discrete time stochastic analysis: basic results 49
    6.1   Basic notions: stochastic basis, predictability
           and martingales 49
    6.2   Martingales on finite time interval 55
    6.3   Martingales on infinite time interval 59

7	**Discrete time stochastic analysis: further results and applications**	**65**
	7.1 Limiting behavior of martingales with statistical applications	65
	7.2 Martingales and absolute continuity of measures. Discrete time Girsanov theorem and its financial application	74
	7.3 Asymptotic martingales and other extensions of martingales	76
8	**Elements of classical theory of stochastic processes**	**81**
	8.1 Stochastic processes: definitions, properties and classical examples	81
	8.2 Stochastic integrals with respect to a Wiener process	91
	8.3 The Ito processes: Formula of changing of variables, theorem of Girsanov, representation of martingales	98
9	**Stochastic differential equations, diffusion processes and their applications**	**107**
	9.1 Stochastic differential equations	107
	9.2 Diffusion processes and their connection with SDEs and PDEs	117
	9.3 Applications to Mathematical Finance and Statistics of Random Processes	126
	9.4 Controlled diffusion processes and applications to option pricing	132
10	**General theory of stochastic processes under "usual conditions"**	**139**
	10.1 Basic elements of martingale theory	139
	10.2 Extension of martingale theory by localization of stochastic processes	151
	10.3 On stochastic calculus for semimartingales	159
	10.4 The Doob-Meyer decomposition: proof and related remarks	169
11	**General theory of stochastic processes in applications**	**175**
	11.1 Stochastic mathematical finance	175
	11.2 Stochastic Regression Analysis	182
12	**Supplementary problems**	**189**
References		**203**
Index		**205**

Acronyms and Notation

$(\Omega, \mathcal{F}, P)$	Probability space
$(\mathcal{F}_n)_{n=0,1...}$	Filtration
$(\Omega, \mathcal{F}, (\mathcal{F}_n), P)$	Stochastic basis
R^d	d-dimensional Euclidian space
$\mathcal{B}(R^d)$	Borel σ-algebra on R^d
$(R^d, \mathcal{B}(R^d))$	Borel space
$\mathcal{A}^+$	Class of increasing integrable processes
$\mathcal{A}$	Class of processes with integrable variation
$\mathcal{A}^+_{loc}$	Class of increasing locally integrable processes
$\mathcal{A}_{loc}$	Class of processes with locally integrable variation
$\mathcal{V}$	Class of processes with finite variation
L^P	Space of random variables with finite p-moment
O	Optional σ-algebra
$\mathcal{P}$	Predictable σ-algebra
$\mathcal{M}$	Set of uniformly integrable martingales
$\mathcal{M}^2$	Set of square integrable martingales
$\mathcal{M}_{loc}$	Set of local martingales
$\mathcal{M}^2_{loc}$	Set of locally square integrable martingales
$\mathbb{E}X$	Expected value of random variable X
$Var(X)$	Variance of X
$\mathbb{E}(X\|A)$	Conditional expected value of random variable X with respect to σ-algebra A
$X_n \xrightarrow{P} X$	Convergence in probability
$X_n \to X(a.s.)$	Convergence almost surely
$X_n \xrightarrow{d} X$	Convergence in distribution
$X_n \xrightarrow{L^p} X$	Convergence in L^P space
$X_n \xrightarrow{w} X$	Weak convergence

$Law_P(X)$	Distribution of X w.r. to P
$Law(X, P)$	Distribution of X with respect to measure P
$\tilde{P} \ll P$	Absolute continuity of measure $\tilde{P}$ with respect to P
$\tilde{P} \sim P$	Equivalence of measures $\tilde{P}$ and P
$\binom{n}{k} = C_n^k$	Number of combinations from n by k
$\langle X, Y \rangle$	Joined quadratic characteristic of X and Y
$[X, Y]$	Joined quadratic bracket X and Y
$[\![\tau, \sigma]\!]$	Stochastic integral with finite limits τ and σ
$[\![\tau]\!]$	Graph of τ
$\mathcal{V}^+$	Class of increasing processes
$\langle X, X \rangle$	Quadratic characteristic (compensator) of X
$\mathrm{conv}(f_1, f_2, \ldots)$	Linear convolution of vectors $f_1, f_2, \ldots$
$A \cup B$	Union of sets A and B
$A \cap B$	Intersection of sets A and B
A^c	Complement of set A
$P(A)$	Probability of set A
$\varepsilon(X)$	Stochastic exponent of X
$P(A|\mathcal{A})$	Conditional probability of A w.r. to σ-algebra $\mathcal{A}$
$\mathrm{cov}(X, Y)$	Covariance of random variables X and Y
$a \wedge b$	$\mathrm{Min}(a, b)$
$a \vee b$	$\mathrm{Max}(a, b)$
a^+	$\mathrm{Max}(a, 0)$
a^-	$\mathrm{Max}(-a, 0)$
$\|\cdot\|_P$	Norm in space L^P
$l_{X(t,a)}$	Local time of W at level a during $[0, t]$
$\{\omega : X_t \to\}$	Set of ω for which there exists a finite limit $\lim_{t \to \infty} X_t(\omega)$

Chapter 1
Probabilistic Foundations

Abstract In the first chapter, Kolmogorov's axioms of the theory of probability are presented. The choice of the system of the axioms is explained in context of the Caratheodory theorem. Different probability spaces as well as probability distributions are introduced too. The famous Kolmogorov consistency theorem is formulated, and as a consequence, a brief construction of the Wiener measure is shown (see [1], [6], [7], [10], [15], [19], [40], and [45]).

1.1 Classical theory and the Kolmogorov axiomatics

Classical theory of probability deals with a *finite probability space*, i.e. it creates a mathematical model for any *random (stochastic) experiment* with *finite* number of possible outcomes (elementary events, results) $\omega \in \Omega$, $|\Omega| = N < \infty$, where Ω is the whole set of outcomes.

Any other events $A, B,$ and C are simply called *events*. We cannot be certain of the event A under consideration before a random experiment. Nevertheless, we expect to have a *quantitative measure* of such a possibility. In classical probability, the problem of finding this measure can be solved perfectly.

Denote $\omega_1, ..., \omega_N$ elementary outcomes and define $p_1, ..., p_N$ corresponding possibilities for them. Any other and more complicated event A can be reconstructed with the help of $\omega_1, ..., \omega_N$. Hence, to determine a measure of possibility for event A, one can put

$$P(A) = \sum_{i:\omega_i \in A} p_i.$$

The situation is *different if Ω is not finite*. Consider the well-known random experiment called *Infinite coin trial* with outcomes H(head) and T(tail), and with the *probability of success* (H) p, $p \in (0, 1)$.

It is quite clear that in this case the set Ω can be chosen as follows

$$\Omega = \{\omega : \omega = (a_1, a_2, ...),\ a_i = 0\ or\ 1\}.$$

© The Author(s), under exclusive license to Springer Nature Switzerland AG 2023
A. Melnikov, *A Course of Stochastic Analysis*, CMS/CAIMS Books in Mathematics 6,
https://doi.org/10.1007/978-3-031-25326-3_1

The natural question arises: How big is this set? Mathematically speaking, we should find the power $|\Omega|$ of this set.

It is well-known that any real number $a \in [0, 1)$ can be represented uniquely in the form

$$a = \frac{a_1}{2} + \frac{a_2}{2^2} + ...,$$

this representation provides a one-to-one correspondence between $[0, 1)$ and Ω:

$$a \in [0, 1) \Leftrightarrow \omega = (a_1, a_2, ...) \in \Omega.$$

If we follow to the classical probability theory we should find the way to define the *probability for any elementary event ω*. It is equivalent to a random choice of a point from the semi-interval $[0, 1)$. Taking into account a symmetry of such trials, we must conclude that all elementary events ω are uniformly possible, hence,

$$p = p(\omega) = const = c.$$

Further, we have

$$1 = |[0, 1)| = \text{Probability of } \Omega = \sum_{\omega \in \Omega} p(\omega) = \infty$$

because the set Ω is not countable. So, we must define $p(\omega) = c = 0$, however, this choice is not satisfactory because, for example, for $A = [0, 1/2)$ we have

$$1/2 = \text{Probability of } A = P(A) = \sum_{\omega \in A} p(\omega),$$

and we arrive to a contradiction again.

Based on these considerations, one can conclude that the approach of classical theory of probability must be seriously transformed:

The probability should be well-defined instead of elementary outcomes for a reasonable collection of events of Ω.

The natural candidate for this type of collection is given in the following definition.

Definition 1.1 A collection $\mathcal{A}$ of subsets of given set Ω is called an algebra if the following conditions are fulfilled

1. $\Omega \in \mathcal{A}$,
2. $A, B \in \mathcal{A} \Rightarrow A \cup B \in \mathcal{A}, A \cap B \in \mathcal{A}$,
3. $A \in \mathcal{A} \Rightarrow A^c = \bar{A} = \Omega \setminus A \in \mathcal{A}$.

Using this definition we can realize the idea described above as follows.

Definition 1.2 A function $\mu : \mathcal{A} \to [0, \infty]$ is a **finite-additive measure** on $\mathcal{A}$, if for any disjoint sets $A, B \in \mathcal{A}$:

$$\mu(A \cup B) = \mu(A) + \mu(B).$$

1.1 Classical theory and the Kolmogorov axiomatics

In case of $\mu(\Omega) < \infty$ this measure is called *finite*. In particular case $\mu(\Omega) = 1$ such a measure is called a (finite-additive) *probability measure*. If there is a partition of $\Omega = \cup_{n=1}^{\infty} \Omega_n, \Omega_k \cap \Omega_m = \emptyset, k \neq m, \Omega_n \in \mathcal{A}$ such that $\mu(\Omega_n) < \infty, n = 1, 2, ...$, then μ is σ-*finite*.

Based on these definitions we arrive to the first probability model that will be called a *probability space of the first level*:

$(\Omega, \mathcal{A}, P)$, where $\mathcal{A}$ is an algebra, and P is a probability (finite-additive measure with $P(\Omega) = 1$).

Assume that a finite additive probability P on the probability space of the first level $(\Omega, \mathcal{A}, P)$ is *countably additive* (σ-*additive*). It means that for any sequence of disjoint events $A_n \in \mathcal{A}$ with $\cup_{n=1}^{\infty} A_n \in \mathcal{A}$ the following equality holds:

$$P(\cup_{n=1}^{\infty} A_n) = \sum_{n=1}^{\infty} P(A_n).$$

In this case we call this measure a *probability (probability measure)*. Let us count a *list of natural properties of probability*:

1. $\emptyset \in \mathcal{A} \Rightarrow P(\emptyset) = 0$;
2. $A, B \in \mathcal{A} \Rightarrow P(A \cup B) = P(A) + P(B) - P(A \cap B)$;
3. $A, B \in \mathcal{A}, A \subseteq B \Rightarrow P(A) \leq P(B)$;
4. $A_n \in \mathcal{A}, n = 1, 2, ..., \cup_{n=1}^{\infty} A_n \in \mathcal{A} \Rightarrow P(\cup_{n=1}^{\infty} A_n) \leq \sum_{n=1}^{\infty} P(A_n)$.

Problem 1.1 Prove the properties (1)–(4).

The idea of the proof is shown here in case of property (1): For any $A \in \mathcal{A}$ we have $\emptyset \cap A \in \mathcal{A}$. Further, using the finite-additivity of P and equality $\emptyset \cup \Omega = \Omega$ we obtain

$$1 = P(\emptyset \cup \Omega = \Omega) = P(\emptyset) + P(\Omega) = P(\emptyset) + 1,$$

and, hence, $P(\emptyset) = 0$.

The following theorem contains conditions under which a finite-additive probability measure is countable-additive.

Theorem 1.1 *Let P be a finitely additive set function on the algebra $\mathcal{A}$ with $P(\Omega) = 1$. Then the following sentences are equivalent:*

1. *P is a probability (σ-additive);*
2. *P is continuous from below:*

$$P(\cup_{n=1}^{\infty} A_n) = \lim_{n \to \infty} P(A_n) \; for \; A_1 \subseteq A_2 \subseteq ..., \cup_{n=1}^{\infty} A_n \in \mathcal{A};$$

3. *P is continuous from above:*

$$P(\cap_{n=1}^{\infty} A_n) = \lim_{n \to \infty} P(A_n) \; for \; A_1 \supseteq A_2 \supseteq ..., \cap_{n=1}^{\infty} A_n \in \mathcal{A};$$

4. *P is continuous at $\emptyset$:*

$$\lim_{n\to\infty} P(A_n) = 0 \ for \ A_1 \supseteq A_2 \supseteq ..., \cap_{n=1}^{\infty} A_n = \emptyset.$$

Proof We only show the implication (1) $\Rightarrow$ (2) because the other statements can be proved in a similar way.

Let us represent $\cup_{n=1}^{\infty} A_n = A_1 \cup (A_2 \setminus A_1) \cup (A_3 \setminus A_2) \cup ...$ as a union of disjoint events. Therefore,

$$P(\cup_{n=1}^{\infty} A_n) = P(A_1) + P(A_2) - P(A_1) + ... = \lim_{n\to\infty} P(A_n).$$

The first probability model $(\Omega, \mathcal{A}, P)$ looks very constructive. But in the framework of this model we *cannot operate with events which are combinations of countable numbers of these events*. To avoid the disadvantages of the probability space of the first level we introduce the following notions.

Definition 1.3 A system $\mathcal{F}$ of subsets of Ω is called a σ-**algebra (sigma-algebra)** if $\mathcal{F}$ is an algebra and for any sequence $A_n \in \mathcal{F}$, $n = 1, 2, ...$, their unions and intersections belong to $\mathcal{F}$:

$$A_n \in \mathcal{F} \Rightarrow \cup_{n=1}^{\infty} A_n \ and \ \cap_{n=1}^{\infty} A_n \in \mathcal{F}.$$

The couple $(\Omega, \mathcal{F})$ is called a *measurable space*. Let us investigate *interconnections* of algebras and σ-algebras. *First*, we note that

$$\mathcal{F}_* = \{\emptyset, \Omega\} \ and \ \mathcal{F}^* = \{A, A \subseteq \Omega\} = \{\text{all subsets of } \Omega\}$$

are both algebras and σ-algebras. *Second*, we can fix a set $A \subseteq \Omega$ and define another algebra (and a σ-algebra also) $\mathcal{F}_A = \{A, A^c, \emptyset, \Omega\}$, which is generated by event A. This way of construction of algebras and σ-algebras admits a natural generalization.

Lemma 1.1 *Let $\mathcal{E}$ be a collection of sets in Ω, then there are the smallest algebra and the smallest σ-algebra containing $\mathcal{E}$.*

Idea of the proof. It is clear that $\mathcal{E} \subseteq \mathcal{F}^*$. Therefore, there exist at least one algebra and one σ-algebra with the desirable property. Define $\alpha(\mathcal{E})$ and $\sigma(\mathcal{E})$ as a collection of sets that belong to every algebra (correspondently, σ-algebra) containing $\mathcal{E}$.

Definition 1.4 A collection of subsets of Ω is called a *monotonic class* if for any increasing (decreasing) sequence of events their union (intersection) belong to this class.

Denote $\mu(\mathcal{E})$ a minimal monotonic class containing $\mathcal{E}$. It turns out, *an algebra $\mathcal{A}$ is a σ-algebra if and only if it is a monotonic class.*

Idea of the proof. The direct implication of this statement is clear because σ-algebra is a monotonic class. The inverse implication can be proved as follows. If $(A_n)_{n=1,2,...}$ is an arbitrary sequence of subsets Ω from a monotonic class $\mathcal{A}$, we can define a monotonic class $\mathcal{A}$, and a monotonic sequence $(B_n)_{n=1,2,...}$ with $B_n = \cup_{i=1}^{n} A_i \in \mathcal{A}$, and then $B_n \uparrow \cup_{i=1}^{\infty} A_i \in \mathcal{A}$. Using this statement we can easily prove that $\mu(\mathcal{A}) = \sigma(\mathcal{A})$ if $\mathcal{A}$ is an algebra.

1.1 Classical theory and the Kolmogorov axiomatics

Example 1.1 (Borel space). Let us take $R = R^1 = (-\infty, \infty)$ as Ω, and consider subintervals
$$(a, b] = \{x \in R : a < x \leq b\} \text{ for } -\infty \leq a < b < \infty,$$

Let $\mathcal{A}$ be a system of subsets A of the form:

$$A = \sum_{i=1}^{n} (a_i, b_i] = \cup_{i=1}^{n} (a_i, b_i], \ n = 1, 2, \ldots$$

Including to $\mathcal{A}$ the empty set $\emptyset$, we can see that $\mathcal{A}$ is an algebra. Moreover, we can easily define a measure on $\mathcal{A}: l_0(A) = \sum_{i=1}^{n} (b_i - a_i)$. In particular,

$$l_0((a, b]) = b - a = \text{the length of } (a, b].$$

We note also that $\mathcal{A}$ is not a σ-algebra because $A_n = (0, 1 - 1/n] \in \mathcal{A}$, but the union $\cup_{n=1}^{\infty} A_n = (0, 1) \notin \mathcal{A}$.

Even in the framework of this example the following problem arises:
Determine a minimal σ-algebra $\mathcal{B}(R)$ containing $\mathcal{A}$ and a measure on the measurable space $(R, \mathcal{B}(R))$ which is called the Borel space.

Problem 1.2 Prove that

1. $\mathcal{B}(R)$ is equivalent to a minimal σ-algebra containing all intervals $(a, b]$;
2. $\mathcal{B}(R)$ contains $(a, b), [a, b], \{a\}$.

Existence of $\mathcal{B}(R)$ is obvious, but an extension of l_0 from $\mathcal{A}$ to $\mathcal{B}(R)$ is a real mathematical problem, and its solution is the famous *Lebesgue measure l* on $(R, \mathcal{B}(R))$. To avoid real technical difficulties we would like to get a solution of this problem from the fundamental *theorem of Caratheodory*:

Theorem 1.2 *If Ω is a space with an algebra $\mathcal{A}$ and with a σ-additive measure μ_0 on $(\Omega, \mathcal{A})$, then there exists a unique measure μ on $(\Omega, \sigma(\mathcal{A}))$ such that*

$$\mu(A) = \mu_0(A) \text{ for all } A \in \mathcal{A}.$$

Based on all these preparations and theorems 1.1 and 1.2 one can arrive to the following natural *axioms proposed by Kolmogorov* for definition of the *probability space of the second level* (or simply, *probability space*):

An ordered triple $(\Omega, \mathcal{F}, P)$ with Ω as a set of elements ω, $\mathcal{F}$ as a σ-algebra on Ω, P as a probability measure (probability) on $\mathcal{F}$.

To go further we give some other examples of measurable spaces playing an important role in many our further constructions and statements.

Example 1.2 $(R^d, \mathcal{B}(R^d))$ is a multidimensional $(d \geq 1)$ Borel space.

Take a set $I = \prod_{i=1}^{d} I_i$, $I_i = (a_i, b_i]$, $i = 1, \ldots, d$, which is called a *rectangle with sides I_i*. The smallest σ-algebra containing all such sets is denoted $\mathcal{B}(R^d)$.

Instead of rectangels I, one can consider the sets $B = \prod_{i=1}^{d} B_i$, $B_i \in \mathcal{B}(R^1)$, $i = 1, 2, ..., d$. The minimal σ-algebra containing all such sets is

$$\mathcal{B}(R^d) = \underbrace{\mathcal{B}(R^1) \otimes ... \otimes \mathcal{B}(R^1)}_{d}.$$

Problem 1.3 Using an induction method, prove the above equality.

Example 1.3 Denote

$$R^\infty = \{x = (x_1, x_2, ...), -\infty < x_i < +\infty\},$$

$$C(I) = C(I_1 \times ... \times I_n) = \{x : x = (x_1, x_2, ...), x_1 \in I_1, ..., x_n \in I_n\}.$$

It is clear that the cylinders C satisfy the *consistency property*

$$C(I_1 \times ... \times I_n) = C(I_1 \times ... \times I_n \times R).$$

In this case, the system of the cylinders is an algebra. The smallest σ-algebra containing all cylinders is $\mathcal{B}(R^\infty)$, and we arrive to a measurable space $(R^\infty, \mathcal{B}(R^\infty))$.

Example 1.4 For $T = [0, \infty)$ we define R^T the space of all real-valued functions x, and $C_{t_1,...,t_n}(I_1 \times ... \times I_n) = \{x : x_{t_1} \in I_1, ..., x_{t_n} \in I_n\}$. The smallest σ-algebra containing all such cylinders is $\mathcal{B}(R^T)$. So, we arrive to the measurable space $(R^T, \mathcal{B}(R^T))$. Let us note a *structural property* of this space: For any $A \in \mathcal{B}(R^T)$ there exist $t_1 < t_2 < ...$ and a Borel set $B \in \mathcal{B}(R^\infty)$ such that

$$A = \{x : (x_{t_1}, x_{t_2}, ...) \in B\}.$$

1.2 Probabilistic distributions and the Kolmogorov consistency theorem

After these theoretical considerations and the examples given above, we can put the problem:

How to construct a probability measure on given measurable space?

In the framework of Example 1.1 one can take a Borel set $A \in \mathcal{B}(R)$ in the form

$$A = (-\infty, x], \, x \in R^1,$$

and define the function

$$F(x) = P((-\infty, x]),$$

where P is a probability measure on the Borel space.

We can note that the function $F = F(x)$ satisfies the following obvious properties

1. F is non-decreasing,

1.2 Probabilistic distributions and the Kolmogorov consistency theorem

2. $F(-\infty) = \lim_{x \to -\infty} F(x) = 0$,
   $F(+\infty) = \lim_{x \to +\infty} F(x) = 1$,
3. F is a right-continuous function with finite left-limits.

Such a function, i.e. satisfying (1)-(3), is called a *distribution function*.

It is now clear that the problem of a construction of the probability can be solved if we have a distribution function F. In this case we can define a finite additive set function P_0 on algebra $\mathcal{A}$:

$$P_0(A) = \sum_{i=1}^n (F(b_i) - F(a_i)),$$

$$A = \sum_{i=1}^n (a_i, b_i] \in \mathcal{A}.$$

We can prove that P_0 is σ-additive, and according to the Caratheodory theorem P_0 admits a unique extension P to the whole space $(R^1, \mathcal{B}(R^1))$.

In particular, we can consider the measurable space $([0, 1], \mathcal{B}([0, 1]))$ and the distribution function

$$F(x) = \begin{cases} 0, & x < 0, \\ x, & x \in [0, 1], \\ 1, & x > 1. \end{cases}$$

Then the set function $l((a, b]) = F(b) - F(a) = b - a$ is the length of the interval $(a, b]$ presents the Lebesgue measure on this space.

Further, starting at the distribution function $F = F(x)$ which is a piecewise constant with change in values at points $x_1, x_2, \ldots, \Delta F(x_i) > 0$, we define

$$p_i = P(\{x_i\}) = \Delta F(x_i) > 0, \quad \sum_{i=1}^\infty P(\{x_i\}) = 1.$$

This is a measure concentrated at $x_1, x_2, \ldots$, and usually the sequence $(p_1, p_2, \ldots)$ is called a *discrete distribution*. Sometimes, the table

$$\begin{pmatrix} p_1 \ldots p_n \ldots \\ x_1 \ldots x_n \ldots \end{pmatrix}$$

is also called a *discrete distribution*. The corresponding distribution function F is also called *discrete*. Below we have the most well-known discrete distributions:

1. Uniform discrete distribution: $\begin{pmatrix} x_1 \ldots x_N \\ 1/N \ldots 1/N \end{pmatrix}$;
2. Bernoulli distribution: $\begin{pmatrix} x_1 & x_2 \\ p_1 & p_2 \end{pmatrix}$, $p_1 + p_2 = 1$;
3. Binomial distribution with parameter p: $\begin{pmatrix} 0 \ldots i \ldots n \\ \binom{n}{i} p^i (1-p)^{n-i} \end{pmatrix}$, $p \in (0, 1)$;

4. Poisson distribution with parameter λ: $\begin{pmatrix} 0 \ldots i \ldots n \ldots \\ \frac{\lambda^i}{i!} e^{-\lambda} \end{pmatrix}, \lambda > 0.$

If the distribution function F admits the integral representation

$$F(x) = \int_{-\infty}^{x} f(y) dy,$$

where a non-negative function (*density*) $f(y)$ satisfies the integral condition $\int_{-\infty}^{\infty} f(y) dy = 1$, then the distribution and its distribution function are *absolute continuous*.

Let us count the most important *absolute continuous distributions* as examples:

1. Uniform distribution on $[a, b]$:

$$f(y) = \frac{1}{b-a}, \quad y \in [a, b];$$

2. Normal distribution with parameters μ and σ^2 ($N \sim (\mu, \sigma^2)$) :

$$f(y) = (2\pi\sigma^2)^{-1/2} exp\left\{-\frac{(y-\mu)^2}{2\sigma^2}\right\}, \quad \mu \in R^1, \sigma > 0, y \in R^1;$$

3. Gamma distribution:

$$f(y) = \frac{y^{\alpha-1} e^{-y/\beta}}{\Gamma(\alpha)\beta^\alpha}, \quad y \geq 0, \alpha > 0, \beta > 0;$$

In particular, for $\beta = 1/\lambda$, and $\alpha = 1$, we get an *exponential distribution* with parameter

$$\lambda \geq 0: \quad f(y) = \lambda e^{-\lambda y}, \quad y \geq 0;$$

for $\alpha = n/2$, and $\beta = 2$, we get *the Chi-squared distribution* with

$$f(y) = 2^{-n/2} y^{n/2-1} e^{-y/2} / \Gamma(n/2), \quad y \geq 0, n = 1, 2, \ldots;$$

4. Student distribution (t-distribution):

$$f(y) = \frac{\Gamma((n+1)/2)}{(n\pi)^{1/2}\Gamma(n/2)} (1 + y^2/n)^{-(n+1)/2}, \quad y \in R^1, n = 1, 2, \ldots;$$

5. Cauchy distribution with parameter θ :

$$f(y) = \frac{\theta}{\pi(y^2 + \theta^2)}, \quad y \in R^1, \theta > 0.$$

Besides discrete and absolute continuous distribution functions (measures, distributions) there are distribution functions which are continuous, but the *set of their points of increasing has the Lebesgue measure zero*. These types of distributions (measures) are called *singular*.

1.2 Probabilistic distributions and the Kolmogorov consistency theorem

Let us explain it here with the help of the famous *Cantor function*. We define the functions

$$F_1(x) = \begin{cases} \text{linear function}, & \text{between } (0,0) \text{ and } (1/3, 1/2), \\ 1/2, & x \in [1/3, 2/3], \\ \text{linear function}, & \text{between } (2/3, 1/2) \text{ and } (1,1), \ x \in [2/3, 1], \end{cases}$$

$$F_2(x) = \begin{cases} \text{linear function}, & \text{between } (0,0) \text{ and } (1/9, 1/4), \ x \in [0, 1/9], \\ 1/4, & x \in [1/9, 2/9], \\ \text{linear function}, & \text{between } (2/9, 1/4) \text{ and } (1/3, 1/2), \ x \in (2/9, 1/3), \\ 1/2, & x \in [1/3, 2/3], \\ \text{On } (2/3, 1) & \text{is similar to the interval } (0, 1/3), \end{cases}$$

and so on. The sequence (F_n) converges to a non-decreasing continuous function $F_C(x)$, called the *Cantor function*. We can calculate the length of intervals on which $F_C(x)$ is constant and find

$$\frac{1}{3} + \frac{2}{9} + \frac{4}{27} + \ldots = \frac{1}{3}\sum_{n=0}^{\infty}\left(\frac{2}{3}\right)^n = 1.$$

So, the Lebesgue measure $l(\mathcal{N}) = 0$, where $\mathcal{N}$ is the set of points of increasing of F_C.

Denote μ the measure, generated by the Cantor function F_C and find that $\mu(\mathcal{N}) = 1$ because $\mu(\mathcal{N}) = F_C(1) = 1$. It means that μ and l are *singular*, and this fact is denoted as $l \perp \mu$.

It is possible to give a *general description* of arbitrary distribution functions F:

$$F(x) = \alpha_1 F^1(x) + \alpha_2 F^2(x) + \alpha_3 F^3(x),$$

where $\alpha_i \geq 0$, $i = 1, 2,$ and 3, $\sum_{i=1}^{3} \alpha_i = 1$, F^1 is discrete, F^2 is absolutely continuous, and F^3 is singular.

Let us pay our attention to a d-dimensional Borel space $(R^d, \mathcal{B}(R^d))$. If P is a probability measure on this space, we can define the d-dimensional distribution function:

$$F_d(x_1, \ldots, x_d) = P((-\infty, x_1] \times \ldots \times (-\infty, x_d]) = P((-\infty, x]).$$

Problem 1.4 Prove that

1. $F_d(+\infty, \ldots, +\infty) = 1$;
2. $F_d(x_1, \ldots, x_d) \to 0$ if at least one of $x_1, \ldots, x_d$ converges to $-\infty$;
3. Define operator (for $i = 1, \ldots, d$, $a_i < b_i$)

$$\Delta_{a_i, b_i} F_d(x_1, \ldots, x_d) = F_d(x_1, \ldots, b_i, \ldots, x_d) - F_d(x_1, \ldots, a_i, \ldots, x_d),$$

then

$$\Delta_{a_1,b_1}\ldots\Delta_{a_d,b_d} F_d(x_1,\ldots,x_d) = P((a,b]),$$

where $(a,b] = (a_1,b_1] \times \ldots \times (a_d,b_d]$.

In fact, there is a one-to-one correspondence between d-dimensional distribution functions and probabilities on the space $(R^d, \mathcal{B}(R^d))$.

Let us give *some examples of multidimensional distribution functions*:

1. If $F^1,\ldots,F^d$ are the distribution functions on R^1, then $F_d(x_1\ldots x_d) = \prod_{i=1}^d F^i(x_i)$ defines a d-dimensional distribution function.
2. If
$$F^i(x_i) = \begin{cases} 0, & x_i < 0, \\ x_i, & 0 \le x_i \le 1, \\ 1, & x_i > 1, \end{cases}$$

then $F_d(x_1,\ldots,x_d) = x_1\ldots x_d$ corresponds to the d-dimensional Lebesgue measure.

3. A d-dimensional absolute continuous distribution function $F_d(x_1\ldots x_d)$ is defined by the same manner as in the real line case:
$$F_d(x_1,\ldots,x_d) = \int_{-\infty}^{x_1}\ldots\int_{-\infty}^{x_d} f_d(y_1\ldots y_d) dy_1\ldots dy_d,$$

where f_d is the density function, i.e. $f_d \ge 0$ and $\int_{-\infty}^{\infty}\ldots\int_{-\infty}^{\infty} f_d(y_1\ldots y_d) dy_1\ldots dy_d = 1$.

The most important case is the *multidimensional Normal distribution* function. It is defined by the density

$$f_d(x_1\ldots x_d) = \frac{(\det A)^{1/2}}{(2\pi)^{d/2}} \exp\left\{-\frac{1}{2}\sum_{1 \le i,j \le d} a_{ij}(x_i - \mu_i)(x_j - \mu_j)\right\},$$

where $A = (a_{ij})$ is the inverse matrix of a symmetric positive definite matrix B.

In particular, in case $d = 2$, we have

$$f_2(x_1,x_2) = \frac{1}{2\pi\sigma_1\sigma_2(1-\rho^2)^{1/2}} \exp\left\{-\frac{1}{2(1-\rho^2)}\left[\frac{(x_1-\mu_1)^2}{\sigma_1^2} - 2\rho\frac{(x_1-\mu_1)(x_2-\mu_2)}{\sigma_1\sigma_2} + \frac{(x_2-\mu_2)^2}{\sigma_2^2}\right]\right\}$$

where $\sigma_1,\sigma_2 > 0$, $\mu_1,\mu_2 \in R^1$, $|\rho| < 1$.

Now we need to consider the case of space $(R^\infty, \mathcal{B}(R^\infty))$. Let us take a set $B \in \mathcal{B}(R^n)$ and consider a cylinder

$$C_n(B) = \{x \in R^\infty : (x_1,\ldots,x_n) \in B\}.$$

If P is a probability measure on $(R^\infty, \mathcal{B}(R^\infty))$, we can define $P_n(B) = P(C_n(B))$, $n = 1, 2, \ldots$ a sequence of probability measures on space $(R^n, \mathcal{B}(R^n))$. By construction we have

$$P_{n+1}(B \times R^1) = P_n(B),$$

1.2 Probabilistic distributions and the Kolmogorov consistency theorem

which is called the *consistency condition*. Now we can formulate the *famous theorem of Kolmogorov* which is fundamental for foundations of probability theory.

Theorem 1.3 *(Consistency theorem of Kolmogorov). Let $(P_n)_{n=1,2,...}$ be a system of probability measures on $(R^n, \mathcal{B}(R^n))$ correspondingly, $n = 1, 2, \ldots$, satisfying the consistency property. Then there exists a unique probability measure P on $(R^\infty, \mathcal{B}(R^\infty))$ such that*

$$P(C_n(B)) = P_n(B).$$

We can give an example how to construct this sequence (P_n). To do this we start with the sequence of 1-dimensional distribution functions $(F^n(x))_{n=1,2,...}$, $x \in R^1$. Further, we construct another sequence of distribution functions as follows

$$F_1(x_1) = F^1(x_1),\ x_1 \in R^1;\quad F_2(x_1, x_2) = F^1(x_1) \cdot F^2(x_2),\ x_1, x_2 \in R^1;\ etc.$$

Denote $P_1, P_2, \ldots$ probability measures on $(R^1, \mathcal{B}(R^1))$, $(R^2, \mathcal{B}(R^2)), \ldots$, which correspond to the distribution functions $F_1, F_2, \ldots$.

We can also consider the space $(R^{[0,\infty)}, \mathcal{B}(R^{[0,\infty)}))$ for which one can state a *version of consistency theorem* in the same way as before. Using this theorem one can construct an extremely important measure, called the *Wiener measure*.

Denote $(\phi_t(y|x))_{t\geq 0}$ a family of Normal (Gaussian) densities of y for a fixed x :

$$\phi_t(y|x) = \frac{1}{(2\pi t)^{1/2}} \exp\left\{-\frac{(y-x)^2}{2t}\right\}, y \in R^1.$$

For each $t_1 < t_2 < \ldots < t_n$ and $B = \prod_{i=1}^n I_i$, $I_i = (a_i, b_i]$, $a_i < b_i$, we define the measure

$$P_{(t_1,t_2,\ldots,t_n)}(B) = \int_{I_1} \ldots \int_{I_n} \phi_{t_1}(x_1|0)\phi_{t_2-t_1}(x_2|x_1)\ldots\phi_{t_n-t_{n-1}}(x_n|x_{n-1})dx_1\ldots dx_n.$$

Furthermore, the measure P on cylinder sets can be defined as follows:

$$P(C_{t_1\ldots t_n}(I_1 \times \ldots \times I_n)) = P_{(t_1\ldots t_n)}(I_1 \times \ldots \times I_n),$$

The family of measures $(P_{(t_1\ldots t_n)})_{n=1,2,...}$ is consistent. Hence, according to Theorem 1.3 the measure P can be extended from cylinder sets to the whole space $(R^{[0,\infty)}, \mathcal{B}(R^{[0,\infty)}))$.

Chapter 2
Random variables and their quantitative characteristics

Abstract In the second chapter random variables are introduced and investigated in the framework of axiomatic of Kolmogorov. It is shown a connection of probability distributions and distributions of random variables as well as their distribution functions. The notion of the Lebesgue integral is given in context of definition of moments of random variables (see [1], [7], [10], [15], [19], [21], [40], and [45]).

2.1 Distributions of random variables

P. L. Chebyshev was the first who introduced the notion of random variables as functions of elementary outcomes in the theory of probability. Here we develop this topic in the framework of probability spaces of the second level.

Definition 2.1 Let $(\Omega, \mathcal{F}, P)$ be a probability space and $(R^1, \mathcal{B}(R^1))$ be a Borel space. Consider a mapping
$$X : (\Omega, \mathcal{F}) \to (R^1, \mathcal{B}(R^1)).$$

For a given set $B \in \mathcal{B}(R^1)$ the set $X^{-1}(B) = \{\omega : X(\omega) \in B\}$ is called the inverse image of B.

The mapping X is called a random variable (measurable function) if for any $B \in \mathcal{B}(R^1)$:
$$X^{-1}(B) \in \mathcal{F}. \tag{2.1}$$

Let us note that it is useful to allow to X can take values $\pm\infty$. In this case we call a random variable as an *extended* random variable.

The notion of random variables is very productive as illustrated in the following example.

Example 2.1 Let us fix a set $A \in \mathcal{F}$ and define the *indicator* of A as follows:

$$I_A(\omega) = \begin{cases} 1, & \omega \in A, \\ 0, & \omega \in A^c. \end{cases}$$

Definitely, I_A is a mapping from the space $(\Omega, \mathcal{F})$ to $(R^1, \mathcal{B}(R^1))$. Then for any Borel set B we have

$$\begin{cases} \{\omega : I_A(\omega) \in B\} = A \in \mathcal{F} \text{ if } B \text{ contains 1 only,} \\ \{\omega : I_A(\omega) \in B\} = A^c \in \mathcal{F} \text{ if } B \text{ contains 0 only,} \end{cases}$$

otherwise, we have $\emptyset$ or Ω. Hence, the condition (2.1) is fulfilled for I_A, and this mapping is a random variable taking two values 0 and 1. Therefore, the set of all random events is included to the set of random variables.

Problem 2.1 Assume $A_1, \ldots, A_n$ are disjoint events from $\mathcal{F}$ such that $\cup_{i=1}^{n} A_i = \sum_{i=1}^{n} A_i = \Omega$. Define a mapping X with values $x_1, \ldots, x_n$ on the sets $A_1, \ldots, A_n$, correspondingly. Prove that $X = \sum_{i=1}^{n} x_i I_{A_i}$ is a random variable. Let us call X a *simple* random variable.

Further, if X takes countable numbers of values $x_1, \ldots, x_n, \ldots$ on disjoint sets $A_1, \ldots, A_n, \ldots$ with $\sum_{1}^{\infty} x_i I_{A_i}$, then $X = \sum_{1}^{\infty} x_i I_{A_i}$ is called a *discrete* random variable.

Condition (2.1) can be relaxed with the help of the next useful lemma.

Lemma 2.1 *Assume $\mathcal{E}$ is a system of subsets such that $\sigma(\mathcal{E}) = \mathcal{B}(R^1)$. Then mapping X is a random variable if and only if*

$$X^{-1}(E) \in \mathcal{F} \text{ for all } E \in \mathcal{E}.$$

Proof The direct implication is trivial. To prove the inverse implication we define a system $\mathcal{Y}$ of those Borel sets B for which $X^{-1} \in \mathcal{F}$. Further, we have equalities

$$X^{-1}(\cup_\alpha B_\alpha) = \cup_\alpha (X^{-1}(B_\alpha)),$$
$$X^{-1}(\cap_\alpha B_\alpha) = \cap_\alpha (X^{-1}(B_\alpha)),$$
$$(X^{-1}(B_\alpha))^c = X^{-1}(B_\alpha^c).$$

It follows from here that $\mathcal{Y}$ is a σ-algebra and $\mathcal{E} \subseteq \mathcal{B}(R^1)$. Hence, $\mathcal{B}(R^1) = \sigma(\mathcal{E}) \subseteq \sigma(\mathcal{Y}) = \mathcal{Y}$ and Lemma is proved. □

Based on Lemma 2.1 and the previous description of the Borel space we arrive to the corollary.

Corollary 2.1 *The function X is a random variable $\Leftrightarrow$ $X^{-1}((-\infty, x]) \in \mathcal{F}$ for all $x \in R^1$. This statement is true if we replace $(-\infty, x]$ to $(-\infty, x)$.*

Now we arrive to definition of the most important quantitative characteristic of random variable X.

2.1 Distributions of random variables

Definition 2.2 Probability measure P_X on a Borel space $(R^1, \mathcal{B}(R^1))$ defined as
$$P_X(B) = P(X^{-1}(B)), \ B \in \mathcal{B}(R^1), \tag{2.2}$$
is called a *distribution (probability distribution)* of X.

Applying equality (2.2) to the set $B = (-\infty, x]$, $x \in R^1$, we arrive to the *distribution function (cumulative distribution function)* of X:
$$F_X(x) = P\{\omega : X(\omega) \leq x\}. \tag{2.3}$$

In the previous sections we already studied arbitrary distributions on $(R^1, \mathcal{B}(R^1))$. A very similar illustration can be found in the context of random variables.

Example 2.2 1. For *discrete* random variable X :
$$P_X(B) = \sum_{i:x_i \in B} p(x_i), \ p(x_i) = P(\omega : X(\omega) = x_i) = \Delta F_X(x_i), \ B \in \mathcal{B}(R^1);$$

2. For *continuous* random variable X its distribution function $F_X(x)$ is continuous;
3. For *absolute continuous* random variable X its distribution function $F_X(x)$ has an integral representation with the density $f_X(x)$:
$$F_X(x) = \int_{-\infty}^{x} f_X(y) dy.$$

Usually, any measurable function
$$\varphi : (R^1, \mathcal{B}(R^1)) \to (R^1, \mathcal{B}(R^1)),$$
is called Borelian. Using Borelian functions and a random variable, one can construct many other random variables. This is the essence of the next important lemma.

Lemma 2.2 *Let φ be a Borelian function and X be a random variable on a probability space $(\Omega, \mathcal{F}, P)$, then the mapping*
$$Y = \varphi \circ X = \varphi(X),$$
is a random variable.

Proof For an arbitrary $B \in \mathcal{B}(R^1)$ we have
$$\{\omega : Y(\omega) \in B\} = \{\omega : \varphi(X(\omega)) \in B\} =$$
$$\{\omega : X(\omega) \in \varphi^{-1}(B)\} \in \mathcal{F}.$$
It means that Y is a random variable. In particular, the functions $x^+, x^-, |x|$ etc. are random variables. □

Let us note that there is a necessity to study *multidimensional* random variables or *random vectors*.

The corresponding definition is straightforward. This is a measurable mapping

$$X : (\Omega, \mathcal{F}) \to (R^d, \mathcal{B}(R^d)), \ d \geq 1.$$

In this case we have

$$X(\omega) = (X_1(\omega), \ldots, X_d(\omega)),$$

where $X_i, i = 1, \ldots, d$, are one-dimensional random variables. We can define a distribution function of X as follows

$$F_X(x_1, \ldots, x_d) = P(\omega : X_1(\omega) \leq x_1, \ldots, X_d(\omega) \leq x_d).$$

Definition 2.3 We say that $X_1, \ldots, X_d$ are independent if

$$F_X(x_1, \ldots, x_d) = \prod_{i=1}^{d} F_{X_i}(x_i).$$

The following theorem is very important in providing many probabilistic constructions.

Theorem 2.1 *For any random variable (extended random variable) X there exists a sequence of simple random variables $X_1, X_2, \ldots$, $|X_i| \leq |X|$ such that $X_n(\omega) \to X(\omega)$ for all $\omega \in \Omega$. For a non-negative X such a sequence $(X_n)_{n=1,2,\ldots}$ can be constructed as a non-decreasing sequence.*

Proof Assume $X \geq 0$ and define for $n = 1, 2, \ldots$ simple random variables as follows

$$X_n(\omega) = \sum_{i=1}^{n \cdot 2^n} \frac{i-1}{2^n} I_{i,n}(\omega) + n I_{\{X(\omega) \geq n\}}(\omega), \qquad (2.4)$$

where $I_{i,n} = I_{\{\omega : \frac{i-1}{2^n} \leq X(\omega) < \frac{i}{2^n}\}}$.
The general case follows from here if we represent $X = X^+ - X^-$. □

Problem 2.2 Prove that for (extended) random variables $X_1, X_2, \ldots$ the mappings $\sup_n X_n$, $\inf_n X_n$, $\liminf_n X_n$, $\limsup_n X_n$ are (extended) random variables.
As a *hint* to the corresponding solution we note that for $\sup_n X_n$ we have

$$\{\omega : \sup_n X_n > x\} = \cup_n \{\omega : X_n > x\} \in \mathcal{F}, \ x \in R^1.$$

Theorem 2.2 *Let $(X_n)_{n=1,2,\ldots}$ be a sequence of (extended) random variables. Then $X(\omega) = \lim_{n \to \infty} X_n(\omega)$ is an (extended) random variable.*

Proof Follows from Problem 2.2 and the next equalities: for any $x \in R^1$

2.1 Distributions of random variables

$$\{\omega : X(\omega) < x\} = \{\omega : \lim_{n\to\infty} X_n(\omega) < x\}$$
$$= \{\omega : \limsup_n X_n(\omega) = \liminf_n X_n(\omega)\} \cap \{\omega : \limsup_n X_n(\omega) < x\}$$
$$= \Omega \cap \{\omega : \limsup_n X_n(\omega) < x\}$$
$$= \{\omega : \limsup_n X_n(\omega) < x\} \in \mathcal{F}.$$

Combining previous facts about limiting approximation of given (extended) random variables X and Y with the help of sequences of simple random variables $(X_n)_{n=1,2,...}$ and $(Y_n)_{n=1,2,...}$ correspondingly, we can prove the following convergence properties:

1. $\lim_{n\to\infty}(X_n(\omega) \pm Y_n(\omega)) = X(\omega) \pm Y(\omega), \omega \in \Omega$;
2. $\lim_{n\to\infty}(X_n(\omega) \cdot Y_n(\omega)) = X(\omega) \cdot Y(\omega), \omega \in \Omega$;
3. $\lim_{n\to\infty}(X_n(\omega) \cdot \tilde{Y}_n^{-1}(\omega)) = X(\omega) \cdot Y^{-1}(\omega), \omega \in \Omega$, where $\tilde{Y}_n(\omega) = Y_n(\omega) + \frac{1}{n} I_{\{\omega : Y_n(\omega)=0\}}$.

Now, we focus on the question of how one random variable can be represented by another.

Definition 2.4 For a given random variable X the family of events

$$((\omega : X(\omega) \in B))_{B \in \mathcal{B}(R^1)},$$

is called a σ-*algebra* $\mathcal{F}_X$ generated by X.

Problem 2.3 Prove that $\mathcal{F}_X$ is a σ-algebra.

Theorem 2.3 *Assume a random variable Y is measurable with respect to σ-algebra $\mathcal{F}_X$. Then there exists a Borelian function φ such that*

$$Y = \varphi \circ X.$$

Proof Consider the set of $\mathcal{F}_X$-measurable functions $Y = Y(\omega)$. Denote $\tilde{\mathcal{D}}_X$ the set of $\mathcal{F}_X$-measurable functions of the form $\varphi \circ X$. It is clear that $\tilde{\mathcal{D}}_X \subseteq \mathcal{D}_X$. To prove the inverse inclusion, $\mathcal{D}_X \subseteq \tilde{\mathcal{D}}_X$, consider a set $A \in \mathcal{F}_X$ and $Y(\omega) = I_A(\omega)$.

Note that $A = X^{-1}(B)$ for some $B \in \mathcal{B}(R^1)$, and $Y = I_A(\omega) = I_B(X(\omega)) \in \tilde{\mathcal{D}}_X$.

Further, we consider functions Y of the form $\sum_{i=1}^n c_i I_{A_i}$, where $c_i \in R^1, A_i \in \mathcal{F}_X$, and find that $Y \in \tilde{\mathcal{D}}_X$.

For an arbitrary $\mathcal{F}_X$-measurable function Y we construct a sequence of simple $\mathcal{F}_X$-measurable functions Y_n such that $Y = \lim_n Y_n$. We already know that $Y_n = \varphi_n(X)$ for some Borelian functions φ_n and $\varphi_n(X) \to Y(\omega)$ as $n \to \infty$. Now, we take the set $B = \{x : \lim_{n\to\infty} \varphi_n(x) \text{ exists}\} \in \mathcal{B}(R^1)$ and define a Borelian function

$$\varphi(x) = \begin{cases} \lim_n \varphi_n(x), & x \in B, \\ 0, & x \notin B. \end{cases}$$

Then $Y(\omega) = \lim_{n\to\infty} \varphi_n(X(\omega)) = \varphi(X(\omega))$ for all $\omega \in \Omega$ and we obtain $\tilde{\mathcal{D}}_X \subseteq \mathcal{D}_X$.
□

2.2 Expectations of random variables

In this section we provide a construction of another important and convenient quantitative characteristic of a random variable X on a probability space $(\Omega, \mathcal{F}, P)$.

We start such a construction from the simplest case when

$$X(\omega) = \sum_{i=1}^{n} x_i \cdot I_{A_i}(\omega), \quad \sum_{i=1}^{n} A_i = \Omega, \; x_i \in R^1. \tag{2.5}$$

Definition 2.5 For the simple random variable X with representation (2.5) we define its *expected value (expectation)* $\mathbf{E}X$ as follows

$$\mathbf{E}X = \sum_{i=1}^{n} x_i P(A_i).$$

Problem 2.4 a) For simple random variables $X_1, \ldots, X_m$ and real numbers $a_1, \ldots, a_m$ we have $\mathbf{E} \sum_{i=1}^{m} a_i X_i = \sum_{i=1}^{m} a_i \mathbf{E} X_i$ (linearity of expected values).
b) For two simple random variables $X \leq Y$ we have $\mathbf{E}X \leq \mathbf{E}Y$ (monotonicity of expected values).

Let X be a non-negative random variable. According to Theorem 2.1 one can construct a sequence of simple random variables $X_n^{(\omega)} \to X(\omega)$, $n \to \infty$, for all $\omega \in \Omega$. Further, due to Problem 2.4, the sequence of expected values $(\mathbf{E}X_n)_{n=1,2,\ldots}$ is non-decreasing, and therefore $\lim_{n\to\infty} \mathbf{E}X_n$ does exist (finite number or $+\infty$). It gives a possibility to give the following definition.

Definition 2.6 The expected value (expectation) of a non-negative random variable X is defined as
$$\mathbf{E}X = \lim_{n\to\infty} \mathbf{E}X_n.$$

Problem 2.5 1) If there exists another sequence of simple random variables $\tilde{X}_n(\omega) \uparrow X(\omega)$, $n \uparrow \infty$, $\omega \in \Omega$; then $\lim_{n\to\infty} \mathbf{E}\tilde{X}_n = \mathbf{E}X$.
2) If $0 \leq X(\omega) \leq Y(\omega)$ for all $\omega \in \Omega$, then $\mathbf{E}X \leq \mathbf{E}Y$.

There is a standard way of expected value extensions to random variables taking both positive and negative values. We can represent X as difference of its positive and negative parts: $X = X^+ - X^-$, where as usual $X^+ = max(0, X)$ and $X^- = max(-X, 0)$. In this case, we say that $\mathbf{E}X$ *does exist* if $\mathbf{E}X^+$ or $\mathbf{E}X^- < \infty$ and $\mathbf{E}X = \mathbf{E}X^+ - \mathbf{E}X^-$.

One can say that $\mathbf{E}X$ is *finite* if both $\mathbf{E}X^\pm < \infty$, and $\mathbf{E}|X| = \mathbf{E}(X^+ + X^-) = \mathbf{E}X^+ + \mathbf{E}X^- < \infty$. In this case we also call X *integrable*.

2.2 Expectations of random variables

There is another denotation for expected values:
$$\mathbf{E}X = \int_\Omega X dP.$$

This denotation came from functional analysis, where the notion of expected value of a measurable function X is called the *Lebesgue integral*. As a matter of fact we note that the Lebesgue integral generalizes the Riemann integral at least in two directions:

1. It is constructed on the measurable space without any metric structure.
2. It is well defined for any measurable bounded function X. But construction of the Riemann integral and its existence depends on the power of the set of discontinuity $\mathcal{D}_X$ of X, and the set $\mathcal{D}_X$ should have the Lebesgue measure zero.
   A standard example of function "integrable by Lebesgue and non-integrable by Riemann" is the Dirichlet function on $[0, 1]$:
   $$X(\omega) = \begin{cases} 1 & if\ \omega \in Q, \\ 0 & if\ \omega \in R \setminus Q, \end{cases}$$
   where Q is the set of rational numbers.

Let us formulate several natural properties of expected values as a problem because their proofs are straightforward.

Problem 2.6 1) $\mathbf{E}c \cdot X = c\mathbf{E}X$ if $\mathbf{E}X$ does exist and $c \in R$.

2) $\mathbf{E}X \le \mathbf{E}Y$ if $X(\omega) \le Y(\omega)$, $\omega \in \Omega$, $\mathbf{E}X > -\infty$ or $\mathbf{E}Y < \infty$.
3) $|\mathbf{E}X| \le \mathbf{E}|X|$ if $\mathbf{E}X$ does exist.
4) If $\mathbf{E}X$ exists, then $\mathbf{E}X \cdot I_A$ exists for each $A \in \mathcal{F}$. Moreover, if $\mathbf{E}X$ is finite, then $\mathbf{E}XI_A$ is finite too.
5) If X and Y are non-negative or integrable, then $\mathbf{E}(X + Y) = \mathbf{E}X + \mathbf{E}Y$.

To formulate other important properties of expectations we need the next definition.

Definition 2.7 We say that a property holds *almost surely* (a.s.) if there exists a set $\mathcal{N} \in \mathcal{F}$ such that
  a) the property holds for every element $\omega \in \Omega \setminus \mathcal{N}$,
  b) $P(\mathcal{N}) = 0$.

6) If $X = 0$ (a.s.), then $\mathbf{E}X = 0$.
7) If $X = Y$ (a.s.) and $\mathbf{E}|X| < \infty$, then $\mathbf{E}|Y| < \infty$ and $\mathbf{E}X = \mathbf{E}Y$.
8) If $X \ge 0$ and $\mathbf{E}X = 0$, then $X = 0$ (a.s.).
9) If X and Y are integrable and $\mathbf{E}X \cdot I_A \le \mathbf{E}Y \cdot I_A$ for all $A \in \mathcal{F}$, then $X \le Y$ (a.s.).
   Let us give a *solution* of subproblem 8 in the above list of Problem 2.6:
   Denote $A = \{\omega : X(\omega) > 0\}$ and $A_n = \{\omega : X(\omega) \ge \frac{1}{n}\} \uparrow A$ as $n \to \infty$.
   We note that $0 \le XI_{A_n} \le XI_A$, and hence $\mathbf{E}XI_{A_n} \le \mathbf{E}X = 0$. Next, we have $0 = \mathbf{E}XI_{A_n} \ge \frac{1}{n}P(A_n)$. Hence, $P(A_n) = 0$, $n = 1, 2, \ldots$, and $P(A) = \lim_{n \to \infty} P(A_n) = 0$.

Another useful property is the *formula of change of variables* in the Lebesgue integral. Let X be a random variables and ϕ be a Borelian function which is integrable with respect to the distribution P_X of X. Then for any $B \in \mathcal{B}(R)$ the formula of change of variables is true:

$$\int_B \phi(x)dP_X = \int_{X^{-1}(B)} \phi(X(\omega))dP. \tag{2.6}$$

In particular, the formula (2.6) for $B = R^1$ is reduced to the formula for calculation of expectation of X:

$$\mathbf{E}\phi(x) = \int_\Omega \phi(X)dP = \int_{-\infty}^\infty \phi(x)dP_X = \int_{-\infty}^\infty \phi(x)dF_X, \tag{2.7}$$

where F_X is a distribution function of X and the integral $\int_{-\infty}^\infty \phi(x)dF_X$ is the Lebesgue-Stiltjes integral.

Scheme of the proof of the formula (2.6) is below

Take $\phi(x) = I_C(x)$, $C \in \mathcal{B}(R^1)$, and (2.6) is reduced to $P_X(B \cap C) = P(X^{-1}(B) \cap X^{-1}(C))$ which follows from the definition of P_X and the equality:

$$X^{-1}(B) \cap X^{-1}(C) = X^{-1}(B \cap C).$$

Next steps are obvious: a non-negative simple function ϕ etc.

If $\phi(x) = x^k$, $k = 1, 2, \ldots$, the expectation $\mathbf{E}X^k$ is called the k-th moment of X with the help of its distribution and its distribution function. Suppose $\mathbf{E}X = \mu$ and $\phi(x) = (x - \mu)^k$, then corresponding moments are called *centered moments*. The second centered moment is called the *variance* of X:

$$Var(X) = \mathbf{E}(X - \mu)^2,$$

and it is one of the key measures of the dispersion of values of X around the mean value μ.

The other common measures of this type are

$$skewness = \frac{\mathbf{E}(X - \mu)^3}{(Var(x))^{3/2}}$$

and

$$kurtosis = \frac{\mathbf{E}(X - \mu)^4}{(Var(X))^2}.$$

Problem 2.7 Let X be a normal random variable with parameters μ and σ^2. Find $EX, Var(X), skewness$ and $kurtosis$ of X.

Chapter 3
Expectations and convergence of sequences of random variables

Abstract In the third chapter asymptotic properties of sequences of random variables are studied. Lemma of Fatou and the Lebesgue dominated convergence theorem are presented as permanent technical tools of stochastic analysis. It is also emphasized the role of a uniform integrability condition of families of random variables. Classical probabilistic inequalities of Chebyshev, Jensen and Cauchy-Schwartz are proved. It is shown how these inequalities work to investigate interconnections between different types of convergence of sequences of random variables. In particular, the large numbers law (LNL) is derived for the case of independent identically distributed random variables (see [1], [7], [10], [15], [19], [40], and [45]).

3.1 Limit behavior of sequences of random variables in terms of their expected values

One of the main questions here is how to take the limit under the sign of expectation. The first result exploits a monotonicity assumption. That is why the corresponding claim is called *Monotonicity convergence theorem*.

Theorem 3.1 *Let $(X_n)_{n=1,2,...}$ be a sequence of random variables then we have*

1. *if there are random variables X and Y such that $X_n \geq Y$, $\mathbf{E}Y > -\infty$, $X_n \uparrow X$, $n \uparrow \infty$, then $\mathbf{E}X_n \uparrow \mathbf{E}X$, $n \uparrow \infty$;*
2. *if there are random variables X and Y such that $X_n \leq Y$, $\mathbf{E}Y < \infty$, $X_n \downarrow X$, $n \uparrow \infty$, then $\mathbf{E}X_n \downarrow \mathbf{E}X$, $n \uparrow \infty$.*

Proof Let us only prove the first part of the theorem because the second part can be derived in the same way. Consider only case $Y \geq 0$, and approximate X_i for each $i = 1, 2, \ldots$ by a sequence $(X_i^n)_{n=1,2,...}$ of simple random variables. Define $X^n = \max_{1 \leq i \leq n} X_i^n$ and note that

$$X^{n-1} \leq X^n = \max_{1 \leq i \leq n} X_i^n \leq \max_{1 \leq i \leq n} X_i = X_n.$$

Denote $Z = \lim_{n \to \infty} X^n$ and find that $X = Z$ because $X_i \leq Z \leq X$ for all $i = 1, 2, \ldots$. This inequality follows from $X_i^n \leq X^n \leq X_n, i = 1, \ldots, n$, by taking the limit as $n \to \infty$. The random variables $(X^n)_{n=1,2,\ldots}$ are simple, then

$$\mathbf{E}X = \mathbf{E}Z = \lim_{n \to \infty} \mathbf{E}X^n \leq \lim_{n \to \infty} \mathbf{E}X_n.$$

On the other hand, since $X_n \leq X_{n+1} \leq X$ we have that

$$\mathbf{E}X \geq \lim_{n \to \infty} \mathbf{E}X_n$$

and, hence,

$$\mathbf{E}X = \lim_{n \to \infty} \mathbf{E}X_n.$$

Problem 3.1 Give the proof in general case when $X_n \geq Y$ and $\mathbf{E}Y > -\infty$.

The next theorem is the most exploited result about taking the limit under the expectation sign. It is called the *Fatou Lemma*.

Theorem 3.2 *1. Assume $X_n \geq Y$ and $\mathbf{E}Y > -\infty$, then*

$$\mathbf{E} \liminf_{n \to \infty} X_n \leq \liminf_{n \to \infty} \mathbf{E}X_n.$$

2. Assume $X_n \leq Y$ and $\mathbf{E}Y < \infty$, then

$$\limsup_{n \to \infty} \mathbf{E}X_n \leq \mathbf{E} \limsup_{n \to \infty} X_n.$$

3. Assume $|X_n| \leq Y$ and $\mathbf{E}Y < \infty$, then

$$\mathbf{E} \liminf_{n \to \infty} X_n \leq \liminf_{n \to \infty} \mathbf{E}X_n \leq \limsup_{n \to \infty} \mathbf{E}X_n \leq \mathbf{E} \limsup_{n \to \infty} X_n.$$

Proof In case (1) we define a sequence of random variables $Z_n = \inf_{m \geq n} X_m$ and find that $Z_n \uparrow \liminf_{n \to \infty} X_n = \sup \inf_{m \geq n} X_n$ and $Z_n \geq Y$. Further we have that

$$\liminf_{n \to \infty} X_n = \liminf_{m \geq n} X_m = \lim_{n \to \infty} Z_n,$$

and applying Theorem 3.1 we get the first claim of the Theorem.

The case (2) is derived in a similar way. The case (3) is just a combination of (1) and (2). □

Let us give the following definition of *convergence of almost surely (a.s)*.

Definition 3.1 We say that sequence $(X_n)_{n=1,2,\ldots}$ converges to a random variable X almost surely (a.s.), if

$$P(\omega : X_n(\omega) \xrightarrow[n \to \infty]{} X(\omega)) = 1.$$

3.1 Limit behavior of sequences of random variables in terms of their expected values

We introduce the following denotations in this case: $X_n \xrightarrow[n \to \infty]{a.s.} X$ or $X_n \xrightarrow[n \to \infty]{} X$ (a.s.).

The following natural question arises here:
Under what conditions one can provide convergence of expected values if $X_n \to X (a.s.)$?

The classical result is the *dominated convergence theorem of Lebesgue*.

Theorem 3.3 *Assume that* $X_n \xrightarrow[n \to \infty]{} X$ *(a.s.) and* $|X_n| \leq Y, \mathbf{E}Y < \infty$. *Then* $\mathbf{E}|X| < \infty$, *and* $\mathbf{E}X_n \to \mathbf{E}X$, $n \to \infty$. *Moreover, the sequence* $(X_n)_{n=1,2,...}$ *converges to X in space* L^1, *i.e.,* $\mathbf{E}|X_n - X| \to 0, n \to \infty$.

Proof Using the Fatou Lemma we obtain $\mathbf{E} \liminf_{n \to \infty} X_n = \liminf_{n \to \infty} \mathbf{E}X_n = \limsup_{n \to \infty} \mathbf{E}X_n = \mathbf{E} \limsup_{n \to \infty} X_n = \mathbf{E}X$. It is clear that $|X| \leq Y$, and therefore $\mathbf{E}|X| < \infty$. Taking into account inequality $|X_n - X| \leq 2Y$ and applying the Fatou Lemma again, we obtain the last statement of the theorem. □

Now we introduce the weakest condition to provide the convergence of expected values.

Definition 3.2 A sequence of random variables $(X_n)_{n=1,2,...}$ is *uniformly integrable*, if
$$\sup_n \mathbf{E}|X_n| I_{\{\omega : |X_n| > c\}} \to 0, \ c \to \infty. \tag{3.1}$$

Problem 3.2 If a sequence of $|X_n| \leq Y$, $n = 1, 2, \ldots, \mathbf{E}Y < \infty$, then $(X_n)_{n=1,2,...}$ is uniformly integrable.

Theorem 3.4 *If a sequence* $(X_n)_{n=1,2,...}$ *is uniformly integrable, then*
1. $\mathbf{E} \liminf_{n \to \infty} X_n \leq \liminf_{n \to \infty} \mathbf{E}X_n \leq \limsup_{n \to \infty} \mathbf{E}X_n \leq \mathbf{E} \limsup_{n \to \infty} X_n$.
2. *if also* $X_n \to X$ *(a.s.),* $n \to \infty$, *then* $\mathbf{E}|X| < \infty$ *and* $\mathbf{E}X_n \to \mathbf{E}X$ *and* $\mathbf{E}|X_n - X| \to 0, n \to \infty$.

Proof To prove (1) we take $c > 0$ and find that
$$\mathbf{E}X_n = \mathbf{E}X_n I_{\{X_n < -c\}} + \mathbf{E}X_n I_{\{X_n \geq -c\}}. \tag{3.2}$$

Using (3.1) for $\epsilon > 0$ we obtain
$$\sup_n |\mathbf{E}X_n I_{\{X_n < -c\}}| < \epsilon.$$

Let us apply the Fatou Lemma to the second term of (3.2) and get
$$\liminf_n \mathbf{E}X_n I_{\{X_n \geq -c\}} \geq \mathbf{E} \liminf X_n I_{\{X_n \geq -c\}}. \tag{3.3}$$

Taking into account that
$$X_n I_{\{X_n \geq -c\}} \geq X_n,$$

we obtain from (3.3) the following inequality

$$\liminf_n \mathbf{E}X_n \geq \mathbf{E}\liminf_n X_n - \epsilon.$$

The proof of part (2) is similar to the proof of Theorem 3.3. □

Theorem 3.5 *Assume the sequence of integrable non-negative random variables* $(X_n)_{n=1,2,...}$ *converges to a random variable* X. *Then* $\mathbf{E}X_n \to \mathbf{E}X < \infty$, $n \to \infty$, *if and only if* $(X_n)_{n=1,2,...}$ *is uniformly integrable.*

Proof The inverse implication follows from Theorem 3.4. To prove the direct implication we consider the set $A = \{a \in R^1 : P(\omega : X(\omega) = a) > 0\}$ and find that

$$X_n I_{\{X_n < a\}} \to X I_{\{X < a\}}, \quad n \to \infty,$$

for each $a \in A$ and $(X_n I_{\{X_n < a\}})_{n=1,2,...}$ is uniformly integrable.
By Theorem 3.4 we obtain

$$\mathbf{E}X_n I_{\{X_n < a\}} \to \mathbf{E}X I_{\{X < a\}},$$

for $a \in A$ and hence,

$$\mathbf{E}X_n I_{\{X_n \geq a\}} \to \mathbf{E}X I_{\{X \geq a\}},$$

for $a \notin A$.

For each $\epsilon > 0$ we take $a_0 \notin A$ large enough that

$$\mathbf{E}X I_{\{X \geq a_0\}} < \epsilon/2.$$

Choose N_0 so large to provide inequalities (for all $n \geq N_0$):

$$\mathbf{E}X_n I_{\{X_n \geq a_0\}} \leq \mathbf{E}X I_{\{X \geq a_0\}} + \epsilon/2,$$

and $\mathbf{E}X_n I_{\{X_n \geq a_0\}} \leq \epsilon$. Now we choose $a_1 \geq a_0$ large enough to have the following inequality (for all $n \leq N_0$):

$$\mathbf{E}X_n I_{\{X_n \geq a_1\}} \leq \epsilon.$$

Finally, we obtain

$$\sup_n \mathbf{E}X_n I_{\{X_n \geq a_1\}} \leq \epsilon,$$

which means the condition (3.1).

Let us note the following *necessary condition* of (3.1).

Problem 3.3 If the sequence $(X_n)_{n=1,2,...}$ is uniformly integrable, then

$$\sup_n \mathbf{E}|X_n| < \infty.$$

The following problem provides a convenient *sufficient condition* for (3.1).

Problem 3.4 Let $(X_n)_{n=1,2,...}$ be a sequence of integrable random variables and $G = G(u)$, $u \geq 0$, be a non-negative increasing function such that

3.1 Limit behavior of sequences of random variables in terms of their expected values

$$\lim_{u \to \infty} \frac{G(u)}{u} = \infty, \ \sup_n \mathbf{E} G(|X_n|) < \infty.$$

Then $(X_n)_{n=1,2,\ldots}$ is uniformly integrable.

Hint. Let $\epsilon > 0$ and $G = \sup_n \mathbf{E} G(X_n)$. For large enough c and $t \geq c$, $G(t)/t \geq G/\epsilon$. Then $\mathbf{E}|X_n|I_{\{|X_n|\geq c\}} \leq \frac{\epsilon}{G}\mathbf{E} G(|X_n|)I_{\{|X_n|\geq c\}} \leq \epsilon$.

To conclude our discussion of the notion of uniform integrability, we present another necessary and sufficient condition in the following theorem.

Theorem 3.6 *A sequence of random variables $(X_n)_{n=1,2,\ldots}$ is uniformly integrable if and only if (3.3) is true and*

$$\sup_n \mathbf{E}|X_n|I_A \to 0 \text{ as } P(A) \to 0. \tag{3.4}$$

Proof The direct implication follows from the next considerations. For $c > 0$ we have

$$\mathbf{E}|X_n|I_A = \mathbf{E}|X_n|I_{A \cap \{|X_n|\geq c\}} + \mathbf{E}|X_n|I_{A \cap \{|X_n|<c\}}$$
$$\leq \mathbf{E}|X_n|I_{\{|X_n|\geq c\}} + cP(A).$$

Taking c large enough, we obtain for a fixed $\epsilon > 0$:

$$\sup_n \mathbf{E}|X_n|I_{A \cap \{|X_n|\geq c\}} \leq \epsilon/2. \tag{3.5}$$

From (3.5) with $P(A) \leq \epsilon/(2c)$ we get

$$\sup_n \mathbf{E}|X_n|I_A \leq \epsilon,$$

which means (3.4). To prove the inverse implication we take $P(A) < \delta$ so that $\mathbf{E}|X_n|I_A \leq \epsilon$ uniformly over $n = 1, 2, \ldots$. Let us note that

$$\mathbf{E}|X_n| \geq \mathbf{E}|X_n|I_{A \cap \{|X_n|\geq c\}} \geq cP(|X_n| \geq c).$$

Then

$$\sup_n P(|X_n| \geq c) \leq \frac{1}{c} \sup_n \mathbf{E}|X_n| \to 0, \ c \to \infty.$$

So, we can take as $A = \{|X_n| \geq c\}$ for large enough c and get

$$\sup_n \mathbf{E}|X_n|I_{\{|X_n|\geq c\}} \leq \epsilon,$$

which is equivalent to (3.1). □

3.2 Probabilistic inequalities and interconnections between types of convergence of random variables

We start with some important and well-known probabilistic inequalities.

Chebyshev inequality: For a non-negative random variable X

$$P(\omega : X(\omega) \geq \epsilon) \leq \frac{\mathbf{E}X}{\epsilon}, \ \epsilon > 0.$$

In particular, for arbitrary random variable X with finite $\mathbf{E}X$ and $Var(X)$:

$$P(\omega : X(\omega) \geq \epsilon) \leq \frac{\mathbf{E}X^2}{\epsilon^2} \text{ and}$$

$$P(\omega : |X(\omega) - \mathbf{E}X| \geq \epsilon) \leq \frac{Var(X)}{\epsilon^2}.$$

For the proof we just note that

$$\mathbf{E}X \geq \mathbf{E}XI_{\{X \geq \epsilon\}} \geq \epsilon \mathbf{E}I_{\{X \geq \epsilon\}} = \epsilon P(\omega : X \geq \epsilon).$$

Cauchy-Schwartz inequality: Let X and Y be random variables which are integrable in square: $\mathbf{E}X^2 < \infty$, $\mathbf{E}Y^2 < \infty$, i.e. $X, Y \in L^2$. Then $\mathbf{E}|XY| < \infty$ and

$$\mathbf{E}|XY| \leq (\mathbf{E}X^2 \mathbf{E}Y^2)^{1/2}.$$

Proof Without loss of generality we assume that $\mathbf{E}X^2 > 0$ and $\mathbf{E}Y^2 > 0$. Consider transformed random variables

$$\tilde{X} = X/(\mathbf{E}X^2)^{1/2}, \tilde{Y} = Y/(\mathbf{E}Y^2)^{1/2}.$$

Note that

$$2|\tilde{X}\tilde{Y}| \leq \tilde{X}^2 + \tilde{Y}^2,$$

and obtain

$$2\mathbf{E}|\tilde{X}\tilde{Y}| \leq \mathbf{E}\tilde{X}^2 + \mathbf{E}\tilde{Y}^2 = 2,$$

or $\mathbf{E}|\tilde{X}\tilde{Y}| \leq 1$. □

Jensen inequality: Let a Borelian function $g = g(x)$ be convex downward and $\mathbf{E}g(|X|) < \infty$. Then

$$g(\mathbf{E}X) \leq \mathbf{E}g(X).$$

Proof Let us consider only the case of smooth function $g \in C^2$ with $g'(x) \geq 0$ for all $x \in R^1$. Using the Taylor decomposition at $\mu = \mathbf{E}X$, we have

$$g(x) = g(\mu) + g'(\mu)(x - \mu) + \frac{g''(\theta)(x - \mu)^2}{2}, \qquad (3.6)$$

where θ is between x and μ.

3.2 Probabilistic inequalities and interconnections ...

Putting $x = X$ and taking expectation in (3.6) we get the desirable inequality. □

In context of the Cauchy-Schwartz inequality we want to emphasize one important case when the inequality is transformed to equality.

Theorem 3.7 *Let X and Y be integrable independent random variables, i.e. $F_{XY}(x, y) = F_X(x) \cdot F_Y(y)$. Then $\mathbf{E}|XY| < \infty$ and*

$$\mathbf{E}XY = \mathbf{E}X\mathbf{E}Y. \tag{3.7}$$

Proof We start with non-negative X and Y, constructing sequences $(X_n)_{n=1,2,\ldots}$ and $(Y_n)_{n=1,2,\ldots}$ of discrete random variables such that

$$X_n = \sum_{m=0}^{\infty} \frac{m}{n} I_{\{\frac{m}{n} \leq X(\omega) < \frac{m+1}{n}\}}, \ n = 1, 2, \ldots$$

(Y_n is similar)

$$X_n \leq X, \ Y_n \leq Y, \ |X_n - X| \leq \frac{1}{n}, \ |Y_n - Y| \leq \frac{1}{n}.$$

Since X and Y are integrable we get that $\mathbf{E}X_n \to \mathbf{E}X$, $\mathbf{E}Y_n \to \mathbf{E}Y$, $n \to \infty$, by the Lebesgue dominated convergence theorem. Due to the independence assumption we have

$$\mathbf{E}X_n Y_n = \sum_{m,l} \frac{ml}{n^2} \mathbf{E} I_{\{\frac{m}{n} \leq X(\omega) < \frac{m+1}{n}\}} I_{\{\frac{l}{n} \leq Y(\omega) < \frac{l+1}{n}\}}$$

$$= \sum_{m,l} \frac{ml}{n^2} \mathbf{E} I_{\{\frac{m}{n} \leq X(\omega) < \frac{m+1}{n}\}} \mathbf{E} I_{\{\frac{l}{n} \leq Y(\omega) < \frac{l+1}{n}\}}$$

$$= \mathbf{E}X_n \mathbf{E}Y_n.$$

Let us note that for $n = 1, 2, \ldots$

$$|\mathbf{E}XY - \mathbf{E}X_n Y_n| \leq \mathbf{E}|XY - X_n Y_n|$$
$$\leq \mathbf{E}|X||Y - Y_n| + \mathbf{E}|Y_n||X - X_n|$$
$$\leq \frac{\mathbf{E}X}{n} + \frac{\mathbf{E}(Y + 1/n)}{n} \to 0 \ as \ n \to \infty.$$

Therefore,

$$\mathbf{E}XY = \lim_n \mathbf{E}X_n Y_n = \lim_n \mathbf{E}X_n \lim_n \mathbf{E}Y_n = \mathbf{E}X\mathbf{E}Y < \infty.$$

General case (3.7) can be treated in a similar way if we note equalities: $X = X^+ - X^-$, $Y = Y^+ - Y^-$, $XY = X^+Y^+ - X^-Y^+ - X^+Y^- + X^-Y^-$. □

The brilliant Chebyshev inequality has a lot of applications, and one of the most important corollaries is the Large Numbers Law (LNL).

Theorem 3.8 *Let $(Y_n)_{n=1,2,\ldots}$ be a sequence of identically distributed independent random variables with mean μ and variance σ^2, and $S_n = \sum_{i=1}^{n} Y_i$. Then for any $\epsilon > 0$*

$$P\left(\omega : \left|\frac{S_n}{n} - \mu\right| \geq \epsilon\right) \to 0, \; n \to \infty. \tag{3.8}$$

The proof of (3.8) follows from the Chebyshev inequality and Theorem 3.6. Denoting $\tilde{Y}_i = Y_i - \mu$, $i = 1, 2, \ldots$, we calculate

$$\mathbf{E}\left(\frac{S_n}{n} - \mu\right)^2 = \frac{1}{n^2} \mathbf{E}\left(\sum_{i=1}^{n} \tilde{Y}_i\right)^2$$

$$= \frac{1}{n^2}\left[\mathbf{E}\sum_{i=1}^{n}\tilde{Y}_i^2 + \sum_{i \neq j}\mathbf{E}\tilde{Y}_i\tilde{Y}_j\right]$$

$$= \frac{1}{n^2}\left[n\sigma^2 + \sum_{i \neq j}\mathbf{E}\tilde{Y}_i\tilde{Y}_j\right] = \frac{\sigma^2}{n}.$$

and obtain that

$$P\left(\omega : \left|\frac{S_n}{n} - \mu\right| \geq \epsilon\right) \leq \frac{\mathbf{E}\left(\frac{S_n}{n} - \mu\right)^2}{\epsilon^2} = \frac{\sigma^2}{n\epsilon^2} \to 0, \; n \to \infty.$$

Let us give the following definition.

Definition 3.3 A sequence of random variables $(X_n)_{n=1,2,\ldots}$ converges to a random variable X *in probability*, if for any $\epsilon > 0$

$$P(\omega : |X_n - X| \geq \epsilon) \to 0, \; n \to \infty. \tag{3.9}$$

We denote it as follows $X_n \xrightarrow[n]{P} X$ or $X_n \xrightarrow{P} X$, $n \to \infty$.

Using this definition one can reformulate Theorem 3.8 as convergence in probability of normed sums $X_n = S_n/n$ to the constant μ as $n \to \infty$.

We also introduce two types of convergence of random variables $(X_n)_{n=1,2,\ldots}$ to a random variable X involving expected values.

Definition 3.4 We say that (X_n) converges to X *weakly* (weak convergence), if for any bounded continuous function f :

$$\mathbf{E}f(X_n) \to \mathbf{E}f(X), \; n \to \infty. \tag{3.10}$$

Definition 3.5 Assume X_n, $n = 1, 2, \ldots$ and X belongs to the space L^p with finite p-moment, i.e. $\mathbf{E}|X_n|^p < \infty$, $\mathbf{E}|X|^p < \infty$. We say that X_n converges to X in space

3.2 Probabilistic inequalities and interconnections ...

L^p, if $\mathbf{E}|X_n - X|^p \to 0$, $n \to \infty$. This convergence is denoted as $X_n \xrightarrow[n]{L^p}$ or $X_n \xrightarrow{L^p}$, $n \to \infty$.

Now we are ready to discuss how the convergence (a.s.) (or convergence with probability 1), defined before and these convergences are related to each other.

It is not difficult to construct a sequence $(X_n)_{n=1,2,...}$ of random variables which converges to X for *all* $\omega \in \Omega$, and hence, almost surely.

For example, if X is a random variable and $(a_n)_{n=1,2,...}$ is a sequence of positive numbers converges to zero, then a sequence $X_n(\omega) = (1 - a_n)X(\omega)$ converges to X for all $\omega \in \Omega$.

Further, convergence with probability one provides both convergence in probability and the weak convergence.

If $X_n \to X$ (a.s.), $n \to \infty$, then $|X_n - X| \to 0$ (a.s.), $n \to \infty$, and for $\epsilon > 0$ $I_{\{\omega:|X_n-X|\geq\epsilon\}} \to 0$ (a.s.) as well. Taking expectation we obtain $P(\omega : |X_n - X| \geq \epsilon) = \mathbf{E} I_{\{\omega:|X_n-X|\geq\epsilon\}} \to 0$, $n \to \infty$.

The weak convergence follows due to definition and the Lebesgue dominated convergence theorem.

Convergence in space L^p is failed when the moment of p-th order does not exist. Convergence in probability implies the weak convergence.

To prove it assume that $X_n \xrightarrow[n]{P} X$, f is a bounded continuous function, $|f(x)| \leq c$, $x \in R^1$, $\epsilon > 0$. We can choose a big enough N such that $P(|X| > N) \leq \epsilon/(4c)$. Take $\delta > 0$ so that $|f(x) - f(y)| \leq \epsilon/(2c)$ for $|x| \leq N$ and $|x - y| < \delta$. Then

$$|\mathbf{E}f(X_n) - \mathbf{E}f(X)| \leq \mathbf{E}|f(X_n) - f(X)|I_{\{\omega:|X_n-X|>\delta\}} + \mathbf{E}|f(X_n) - f(X)|I_{\{\omega:|X_n-X|\leq\delta\}}. \tag{3.11}$$

The first term in the right hand side of (3.10) is dominated by $2cP(\omega : |X_n - X| > \delta) \to 0$, $n \to \infty$.

$$\begin{aligned}\mathbf{E}|f(X_n) - f(X)|I_{\{\omega:|X_n-X|\leq\delta\}} &= \mathbf{E}|f(X_n) - f(X)|I_{\{\omega:|X_n-X|\leq\delta\}}I_{\{\omega:|X|\leq N\}} \\ &+ \mathbf{E}|f(X_n) - f(X)|I_{\{\omega:|X_n-X|\leq\delta\}}I_{\{\omega:|X|>N\}} \\ &\leq \epsilon + 2cP(\omega : |X_n - X| > \delta) < 2\epsilon\end{aligned}$$

for a large enough n.

Chapter 4
Weak convergence of sequences of random variables

Abstract The chapter four is devoted in a systematic study of a weak convergence of sequences of random variables. It is shown the equivalence between a weak convergence and convergence in distribution. It is shown that a weak compactness and tightness for families of probability distributions (Prokhorov's theorem) are equivalent. It is discussed a connection between characteristic functions and distributions of random variables. The method of characteristic functions is applied to prove the Central Limit Theorem (CLT) for sums of independent identically distributed random variables (see [6], [15], [21], and [40]).

4.1 Weak convergence and its description in terms of distributions

In the previous section we recognized that *weak convergence* corresponds to the word *"weak"* because all other types of convergence provide the weak one. It turns out this type of convergence plays the most significant role in the theory of probability. This is why we study it here in more details.

We start with the following considerations. Let $(X_n)_{n=1,2,...}$ be a sequence of independent Bernoulli's random variables taking values 1 and 0 with probabilities p and q, $p + q = 1$, correspondingly. In this case, the law of large numbers (Theorem 3.7) is reduced to the following

$$\bar{S}_n = \frac{S_n}{n} \xrightarrow{P} p, \; n \to \infty. \tag{4.1}$$

Let us define distribution functions

$$F_n(x) = P(\omega : \bar{S}_n \leq x) \text{ and } F(x) = \begin{cases} 1, & x \geq p, \\ 0, & x < p, \end{cases} \; x \in R^1.$$

We know also from the previous section that
$$\mathbf{E}f(\bar{S}_n) \to \mathbf{E}f(p), n \to \infty, \quad (4.2)$$

for any bounded continuous function f.

Further, denote P_{F_n} and P_F the distributions on $(R^1, \mathcal{B}(R^1))$ which correspond to F_n and F, respectively. Using formula of change of variables in expectations (2.7), we can rewrite (4.2) in two equivalent forms:
$$\int_{R^1} f(x) dP_{F_n} \to \int_{R^1} f(x) dP_F, \; n \to \infty,$$
$$\int_{R^1} f(x) dF_n \to \int_{R^1} f(x) dF, \; n \to \infty, \quad (4.3)$$

for any bounded continuous function f.

The limit relations (4.3) allow to speak about the weak convergence of distributions P_{F_n} and distribution functions F_n to P_F and F in the sense (4.3).

We also note that
$$F_n(x) \to F(x), \; n \to \infty, \quad (4.4)$$

for all $x \in R^1 \setminus \{p\}$, and hence, the weak convergence can be characterized with the help of distribution functions as their convergence (4.4) generalized in certain way.

Namely, for an arbitrary sequence of random variables $(X_n)_{n=1,2,...}$ and a random variable X with distribution functions $(F_n)_{n=1,2,...}$ and F we may consider a convergence
$$F_n(x) \to F(x), \; n \to \infty,$$

for all points of continuity of F, $x \in R^1$.

Such type of convergence of X_n to X we will call a *convergence in distribution* or convergence in law and denote $X_n \xrightarrow{d} X$, $n \to \infty$.

Theorem 4.1
$$X_n \xrightarrow{w} X, \; n \to \infty \Leftrightarrow X_n \xrightarrow{d} X, n \to \infty.$$

Proof We show the *direct implication* only. The inverse implication is the *Helly's selection principle* application. That is why we omit it here.

For fixed $x \in R^1$ and for each integer $\alpha \geq 1$ we define two bounded continuous functions
$$f_\alpha = f_\alpha(y) = \begin{cases} 1, & y \leq x, \\ \alpha(x-y) + 1, & x < y < x + \alpha^{-1}, \\ 0, & x + \alpha^{-1} \leq y, \end{cases}$$

$$f_\alpha^- = f_\alpha^-(y) = \begin{cases} 1, & x - \alpha^{-1} \geq y, \\ \alpha(x-y), & x - \alpha^{-1} < y \leq x, \\ 0, & x \leq y. \end{cases}$$

Functions f_α and f_α^- admit the limits:

4.1 Weak convergence and its description in terms of distributions

$$\lim_{\alpha \to \infty} f_\alpha(y) = I_{(-\infty, x]}(y) \text{ and } \lim_{\alpha \to \infty} f_\alpha^-(y) = I_{(-\infty, x)}(y). \qquad (4.5)$$

Moreover, applying the Lebesgue dominated convergence theorem we get

$$\lim_{\alpha \to \infty} \mathbf{E} f_\alpha(X) = \mathbf{E} I_{(-\infty, x]}(X) = P(\omega : X(\omega) \leq x) = F(x),$$

$$\lim_{\alpha \to \infty} \mathbf{E} f_\alpha^-(X) = \mathbf{E} I_{(-\infty, x)}(X) = P(\omega : X(\omega) < x) = F(x-). \qquad (4.6)$$

Further, due to $X_n \xrightarrow{w} X$, $n \to \infty$, we have for f_α and f_α^- the limiting relations:

$$\lim_{n \to \infty} \mathbf{E} f_\alpha(X_n) = \mathbf{E} f_\alpha(X),$$

$$\lim_{n \to \infty} \mathbf{E} f_\alpha^-(X_n) = \mathbf{E} f_\alpha^-(X).$$

By construction of f_α and f_α^- we obtain

$$\mathbf{E} f_\alpha^-(X_n) \leq F_n(x) \leq \mathbf{E} f_\alpha(X_n).$$

Hence, for any $\alpha \geq 1$ the following inequalities are true:

$$\mathbf{E} f_\alpha^-(X) \leq \liminf_{n \to \infty} F_n(x) \leq \limsup_{n \to \infty} F_n(x) \leq \mathbf{E} f_\alpha(X). \qquad (4.7)$$

Combining (4.5)-(4.7) we arrive to the inequalities

$$F(x-) \leq \liminf_{n \to \infty} F_n(x) \leq \limsup_{n \to \infty} F_n(x) \leq F(x).$$

which provide convergence $F_n(x) \to F(x)$, $n \to \infty$, if x is a point of continuity of F. $\square$

Let us pay also a brief attention to a characterization of weak convergence of *probability distributions* $(P_n)_{n=1,2,...}$ on space $(R^1, \mathcal{B}(R^1))$.

We start with a very simple observation. Let P and $\tilde{P}$ be two *different* probability measures on $(R^1, \mathcal{B}(R^1))$. Define a sequence of probability measures $(P_n)_{n=1,2,...}$ as follows

$$P_{2n} = P \text{ and } P_{2n+1} = \tilde{P},$$

and we arrive to conclusion that such a sequence $(P_n)_{n=1,2,...}$ *does not converge weakly*.

Another observation: we define a sequence of probability measures such that $P_n(\{n\}) = 1$ and, hence, $P_n(R^1) = 1$. On the other hand, $\lim_{n \to \infty} P_n((a, b]) = 0$ for any $a < b \in R^1$.

What does it mean in terms of distribution functions?

Obviously, P_n has the distribution function

$$F_n(x) = \begin{cases} 1, & x \geq n, \\ 0, & x < n, \ n = 1, 2, \ldots, \end{cases}$$

and for every $x \in R^1$
$$\lim_{n \to \infty} F_n(x) = G(x) = 0.$$

Therefore, the limit above is *not a distribution function*. It means that the set of distribution functions is *not compact*. These observations lead us to the next definitions.

Definition 4.1 A collection of probability measures (P_α) is **relatively compact**, if every sequence of measures from this collection admits a subsequence which converges weakly to a probability measure.

Definition 4.2 A collection of probability measures (P_α) is **tight**, if for every $\epsilon > 0$ there exists a compact set $K \subseteq R^1$ such that
$$\sup_\alpha P_\alpha(R^1 \setminus K) \leq \epsilon.$$

Similar definitions can be reproduced word by word for a collection of distribution functions (F_α).

The classical Prokhorov theorem below speaks us that both these notions are equivalent.

Theorem 4.2 *Let (P_α) be a family of probability measures on $(R^1, \mathcal{B}(R^1))$. Then this family is relatively compact if and only if it is tight.*

Proof We give the proof of the direct implication only. Assume that (P_α) is not tight. Hence, there exist $\epsilon > 0$ such that for any compact $K \subseteq R^1$:
$$\sup_\alpha P_\alpha(R^1 \setminus K) > \epsilon.$$

Taking as K compact intervals $I_n = [-n, n]$, $n = 1, 2, \ldots$, we arrive to a sequence of measures (P_{α_n}) such that $P_{\alpha_n}(R^1 \setminus I_n) > \epsilon$, $n = 1, 2, \ldots$.

Further, since (P_α) is relatively compact, we can select from (P_{α_n}) a subsequence $(P_{\alpha_{n_k}})$ such that $(P_{\alpha_{n_k}}) \xrightarrow{w} \tilde{P}$, $k \to \infty$, where $\tilde{P}$ is a probability measure on $(R^1, \mathcal{B}(R^1))$.

Now we have
$$\epsilon \leq \limsup_{k \to \infty} P_{\alpha_{n_k}}(R^1 \setminus I_n) \leq \tilde{P}(R^1 \setminus I_n),$$

but it is not possible as $n \to \infty$ and $\tilde{P}$ is a probability measure. □

4.2 Weak convergence and Central Limit Theorem

We already know that distribution function F is a key quantitative characteristics of a random variable on given probability space $(\Omega, \mathcal{F}, P)$. Moreover, it is clear from

4.2 Weak convergence and Central Limit Theorem

the Kolmogorov consistency theorem that there always exist such a probability space and a random variable with given distribution. To be more illustrative we would like to describe a particular way how this procedure can be realized.

We assume for simplicity that $F(x)$ is strictly increasing. Then we choose the space $([0, 1], \mathcal{B}([0, 1]), l)$ as a probability space. Let us define the following random variable

$$X(\omega) = \begin{cases} F^{-1}(\omega) & \text{if } \omega \in [0, 1], \ \omega = F(x) \\ 0 \text{ if } \omega < 0 \text{ or } \omega > 1. \end{cases}$$

It is clear that X is $\mathcal{B}([0, 1])$–measurable and if $\omega' \leftrightarrow x'$ and $\omega \leftrightarrow x$, we obtain
$$P(\omega' : X(\omega') \leq x) = P(\omega' : F^{-1}(\omega') \leq x) =$$
$$= P(\omega' : \omega' \leq F(x)) = P(\omega' : \omega' \leq \omega) = F(x),$$

i.e. X has the distribution function $F(x)$.

We also want to emphasize the following additional properties of expectations and variances.

Problem 4.1 Prove that

1. $\mathbf{E} \sum_{i=1}^{n} a_i X_i = \sum_{i=1}^{n} a_i \mathbf{E} X_i$ for integrable random variables $X_1, \ldots, X_n$ and real numbers $a_1, \ldots, a_n$, $n = 1, 2, \ldots$
2. $\mathbf{E} \prod_{i=1}^{n} a_i X_i = \prod_{i=1}^{n} a_i \mathbf{E} X_i$ for independent random variables $X_1, \ldots, X_n$, $n = 1, 2, \ldots$
3. $\mathbf{Var}(aX + b) = a^2 \mathbf{Var}(X)$ for a random variable X with well-defined variance and real numbers a, b
4. $\mathbf{Var}(\sum_{i=1}^{n} a_i X_i) = \sum_{i=1}^{n} a_i^2 \mathbf{Var}(X_i)$ for independent random variables $X_1, \ldots, X_n$ and real numbers $a_1, \ldots, a_n$, $n = 1, 2, \ldots$

Let us pay more attention to normal (or Gaussian) random variables. Denote

$$p(x) = p_{\mu, \sigma^2}(x) = \frac{1}{\sqrt{2\pi}\sigma} \exp\left\{-\frac{(x-\mu)^2}{2\sigma^2}\right\}$$

the density of a normal random variable with parameters μ and σ^2, i.e. $N(\mu, \sigma^2)$. We note the following:

1. If $\mu = 0$ and $\sigma^2 = 1$, then the random variable $Z = N(0, 1)$ is called *standard*.
2. The integral $\int_{-\infty}^{\infty} e^{-x^2/2} dx = \sqrt{2\pi}$ is known as the Poisson integral. Therefore, $\int_{-\infty}^{\infty} p_{0,1}(x) dx = \frac{1}{\sqrt{2\pi}} \sqrt{2\pi} = 1$, which means that $p_{0,1}$ is the density of some probability distribution.
3. $\mathbf{E}Z = \int_{-\infty}^{\infty} x p_{0,1}(x) dx = 0$ because $h(x) = x$ is the odd function, and $\int_{-\infty}^{\infty} y^2 e^{-y^2/2} dy = \sqrt{2\pi}$ for the even function $h(x) = x^2$.

Hence,

$$\mathbf{Var}(Z) = \mathbf{E}Z^2 - (\mathbf{E}Z)^2 = \mathbf{E}Z^2 = \frac{1}{\sqrt{2\pi}} \int_{-\infty}^{\infty} y^2 e^{-y^2/2} dy = 1.$$

4. *Standardization procedure.*
   A random variable $X = N(\mu, \sigma^2) = \sigma Z + \mu$ or, equivalently, $Z = \frac{X-\mu}{\sigma}$. Hence, $\mathbf{E}X = \mathbf{E}(\sigma Z + \mu) = \sigma \mathbf{E}Z + \mu = \mu$ and $\mathbf{Var}(X) = \mathbf{Var}(\sigma Z + \mu) = \sigma^2 \mathbf{Var}(Z) = \sigma^2$, and we get a nice *probabilistic interpretation* of parameters μ and σ^2 as mean and variance of X.

5. Let X be a normal random variable with parameters μ and σ^2. Consider its exponential transformation $Y = e^X$. This equality can be rewritten in a logarithmic form $X = \ln Y$.
   In this case Y is called *log-normal*, and its density has the form

$$l(y) = \frac{1}{\sqrt{2\pi}\sigma} \exp\left(-\frac{(\ln y - \mu)^2}{2\sigma^2}\right) y^{-1}, y \in (0, \infty).$$

Problem 4.2 Denote μ_Y and σ_Y^2 the mean, and variance of Y respectively. Prove that
$$\mu_Y = \mathbf{E}Y = \exp(\mu + \sigma^2/2),$$
$$\sigma_Y^2 = \mathbf{Var}(Y) = \exp(2\mu + \sigma^2)(e^{\sigma^2} - 1).$$

To give an alternative description of moments of a random variable and its distribution function we introduce the notion of characteristic function.

Definition 4.3 Let X be a random variable. Then the function $\phi_X(t) = \mathbf{E}e^{itX}$, $i = \sqrt{-1}$, is called a *characteristic function* of X.

To simplify our further considerations we assume that the distribution function $F_X(x) = F(x)$ admits a density $f_X(x)$. Now we note the following:

First, the characteristics function does exist because of

$$\mathbf{E}|e^{itx}| = \int_{-\infty}^{\infty} |e^{itx}| f_X(x) dx = \int_{-\infty}^{\infty} f_X(x) dx = 1 < \infty.$$

Second, if $\mathbf{E}|X|^m < \infty$, $m = 1, 2, \ldots$, then the characteristic function admits m continuous derivatives such that

$$\mathbf{E}X = i\phi_X'(0), \ \mathbf{E}X^2 = -\phi_X''(0), \ \ldots, \ \mathbf{E}X^m = i^m \phi_X^{(m)}(0). \quad (4.8)$$

Relations (4.8) follow from direct calculations of the corresponding integrals:

$$\phi_X'(t) = \int_{-\infty}^{\infty} ixe^{itx} f_X(x) dx$$

and, hence,

$$\phi_X'(0) = i\int_{-\infty}^{\infty} xe^{i \cdot 0 \cdot x} f_X(x) dx = i\mathbf{E}X \text{ etc.}$$

Example 4.1 1. For the Bernoulli random variable X taking values 1 and 0 with probabilities p and q respectively, $p + q = 1$, we have

4.2 Weak convergence and Central Limit Theorem

$$\phi_X(t) = e^{it \cdot 0} q + e^{it \cdot 1} p = p e^{it} + q.$$

Hence, $\phi'_X(t) = p e^{it} \cdot i$ and $\phi'_X(0) = ip$.

2. For the Poisson random variable with parameter $\lambda > 0$ we have

$$\phi_X(t) = \mathbf{E} e^{itX} = \sum_{m=0}^{\infty} e^{itm} P(\omega : X(\omega) = m) =$$

$$= \sum_{m=0}^{\infty} e^{itm} \frac{\lambda^m}{m!} e^{-\lambda} = \sum_{m=0}^{\infty} \frac{(\lambda e^{it})^m}{m!} e^{-\lambda} =$$

$$= e^{-\lambda} e^{\lambda e^{it}} = e^{\lambda(e^{it} - 1)}$$

and $\phi'_X(0) = \lambda i$.

3. For the normal random variable X with parameters μ and σ^2 we have using standardized random variable Z:

$$\phi_Z(t) = \int_{-\infty}^{\infty} e^{itx} \frac{1}{\sqrt{2\pi}} e^{-x^2/2} dx$$

$$= \int_{-\infty}^{\infty} \frac{\cos(tx)}{\sqrt{2\pi}} e^{-x^2/2} dx + i \int_{-\infty}^{\infty} \frac{\sin(tx)}{\sqrt{2\pi}} e^{-x^2/2} dx$$

$$= \frac{1}{\sqrt{2\pi}} \int_{-\infty}^{\infty} \cos(tx) e^{-x^2/2} dx \qquad (4.9)$$

Differentiating both side of (4.9) and using the integration by parts we obtain

$$\phi'_Z(t) = \frac{1}{\sqrt{2\pi}} \int_{-\infty}^{\infty} (-x \sin(tx)) e^{-x^2/2} dx$$

$$= -\frac{1}{\sqrt{2\pi}} \int_{-\infty}^{\infty} t \cos(tx) e^{-x^2/2} dx = -t \phi_Z(t).$$

Hence,

$$\phi'_Z(t)/\phi Z(t) = -t. \qquad (4.10)$$

Integrating the equation (4.10), we get $\ln \phi_Z(t) = -t^2/2 + c$, $c = const$, and therefore, $\phi_Z(t) = e^c e^{-t^2/2}$ with $e^c = 1$ due to $\phi_Z(0) = 1$. Finally, we have

$$\phi_Z(t) = e^{-t^2/2}.$$

The case of arbitrary $X = N(\mu, \sigma^2)$ is considered with the help of standardization procedure $X = \mu + \sigma Z$. As a result, we arrive to the formula

$$\phi_X(t) = \exp\left(it\mu - \frac{t^2 \sigma^2}{2}\right).$$

Now we want to discuss a correspondence between weak convergence (convergence in distribution, in law) and the convergence of characteristic functions.

There is a simple *sufficient condition* for convergence in distribution. Namely, assume $(f_n)_{n=1,2,...}$ is a sequence of densities of random variables $(X_n)_{n=1,2,...}$ such that $f_n(x) \to f(x)$, $n \to \infty$, implies the convergence of their distribution functions $F_n(x) \to F(x)$, $n \to \infty$. In fact, this result from Calculus is true for any bounded function $h(x)$:

$$\int_{-\infty}^{\infty} h(x) f_n(x) dx \to \int_{-\infty}^{\infty} h(x) f(x) dx, \; n \to \infty.$$

Taking $h(x) = e^{itx}$ in (4.11) we get

$$\phi_{X_n}(t) \to \phi_X(t), \; n \to \infty.$$

The result (4.12) implies the convergence $F_n(x) \to F(x)$, $n \to \infty$, too. Summarizing all these findings we arrive to the following *methodology* to prove the Central Limit Theorem (CLT) of the Theory of Probability.

Theorem 4.3 *Let $(X_n)_{n=1,2,...}$ be a sequence of independent identically distributed (iid) random variables with $\mathbf{E}X_n = \mu$ and $\mathbf{Var}X_n = \sigma^2$. Denote*

$$S_n = \sum_{m=1}^{n} X_m, \; Y_n = \frac{S_n - n\mu}{\sigma \sqrt{n}}.$$

Then

$$Y_n \xrightarrow{d} Y, \; n \to \infty, \tag{4.11}$$

where Y is a standard normal random variable $N(0,1)$.

This is a version of the CLT for iid random variables.

Proof First of all, we note that the characteristic function of the sum of independent random variables is equal to the product of characteristic functions of these random variables. Using this property, we have

$$\phi_{Y_n}(t) = \phi_{\frac{1}{\sigma\sqrt{n}}\sum_{m=1}^n (X_m - \mu)}(t) = \phi_{\sum_{m=1}^n (X_m - \mu)}(t/\sigma\sqrt{n})$$

$$= \prod_{m=1}^{n} \phi_{Z_{X_m - \mu}}(t/\sigma\sqrt{n})$$

$$= (\phi(t/\sigma\sqrt{n}))^n, \; n = 1, 2, \ldots \tag{4.12}$$

Further, $\mathbf{E}(X_m - \mu) = 0$, $\mathbf{E}(X_m - \mu)^2 = \sigma^2$ and

$$\phi'(t) = i\mathbf{E}(X_m - \mu)e^{it(X_m - \mu)}, \; \phi'(0) = 0,$$
$$\phi''(t) = -\mathbf{E}(X_m - \mu)^2 e^{it(X_m - \mu)}, \; \phi''(0) = -\sigma^2.$$

4.2 Weak convergence and Central Limit Theorem

Hence, we can expand ϕ in a Taylor expansion at point $t = 0$ and find that

$$\phi(t) = 1 + 0 - \frac{\sigma^2 t^2}{2} + t^2 \Delta(t),$$

where $\Delta(t) \to 0, t \to 0$. Therefore, using (4.14) we get

$$\phi_{Y_n}(t) = \phi^n(t/\sigma\sqrt{n}) = \exp(n \log \phi(t/\sigma\sqrt{n})) = \exp(n \log(1 - \frac{t^2}{2n} + \frac{t^2}{n\sigma^2}\Delta(t)),$$

and hence $\lim_{n \to \infty} \phi_{Y_n}(t) e^{-t^2/2}$ which is the characteristics function of $N(0, 1)$. □

A more *general version* of the CLT is usually formulated under the *Lindeberg Condition* (**L**).
Let $(X_n)_{n=1,2,...}$ be a sequence of independent random variables with $\mathbf{E} X_n = \mu_n$, $\mathbf{Var} X_n = \sigma_n^2$. We say that the sequence $(X_n)_{n=1,2,...}$ satisfies the condition (**L**) if for any $\epsilon > 0$

$$\left(\sum_{m=1}^n \sigma_m^2\right)^{-1} \sum_{m=1}^n \int_{\left(x:|x-\mu_m| \geq \epsilon (\sum_{m=1}^n \sigma_m^2)^{1/2}\right)} (x - \mu_m)^2 dF_{X_m}(x) \to 0, \ n \to \infty.$$

Under the conditions above

$$\frac{S_n - \mathbf{E} S_n}{(\mathbf{Var} S_n)^{1/2}} \xrightarrow{d} N(0, 1), \ n \to \infty.$$

The *proof* of this version of the CLT is provided by the same method as the proof of Theorem 4.3.

Chapter 5
Absolute continuity of probability measures and conditional expectations

Abstract In this chapter a special attention is devoted to the absolute continuity of measures. It is shown how this notion and the Radon-Nikodym theorem work to define conditional expectations. The list of properties of conditional expectations are given here. In particular, it is emphasized the optimality in the mean-square sense of conditional expectations (see [7], [15], [19], [40], [41] and [45]).

5.1 Absolute continuity of measures and the Radon-Nikodym theorem

Let $(\Omega, \mathcal{F}, P)$ be a probability space and X be a random non-negative variable. Define a set function

$$\tilde{P}(A) = \int_A X \, dP = \mathbf{E} X I_A, \quad A \in \mathcal{F}. \tag{5.1}$$

Let us take $A = \cup_{i=1}^{\infty} A_i$, $A_i \cap A_j = \emptyset, i \neq j$. Using a linearity of expected values we find that $\tilde{P}$ from (5.1) is a *finite-additive measure*. Further, with the help of monotonic convergence theorem we have

$$\tilde{P}(A) = \mathbf{E} X I_A = \mathbf{E} X I_{\cup_{i=1}^{\infty} A_i} =$$
$$= \mathbf{E} \sum_{n=1}^{\infty} X I_{A_n} = \sum_{n=1}^{\infty} \mathbf{E} X I_{A_n} = \sum_{n=1}^{\infty} \tilde{P}(A_n),$$

and therefore $\tilde{P}$ is *countable additive*.

In case of arbitrary random variable we can use the standard representation $X = X^+ - X^-$. Assume that one of expected values $\mathbf{E} X^+$ or $\mathbf{E} X^-$ is finite. In such a case one can define the following *signed-measure*

$$\tilde{P}(A) = \int_A X^+ dP - \int_A X^- dP = \tilde{P}^+(A) - \tilde{P}^-(A), \ A \in \mathcal{F}, \qquad (5.2)$$

that satisfies the next property

$$P(A) = 0 \Rightarrow \tilde{P}(A) = 0. \qquad (5.3)$$

The *proof* of (5.3) is standard. For simple random variable with values $x_1, \ldots, x_n$, $X = \sum_{i=1}^{n} x_i I_{A_i}$ we have

$$\tilde{P}(A) = \mathbf{E} X I_A = \sum_{i=1}^{n} x_i P(A_i \cap A) = 0, \ P(A) = 0.$$

For arbitrary random variable X we monotonically approximate X by a sequence of simple random variables $X_n \uparrow X$, $n \to \infty$, and take a limit according to monotonic convergence theorem:

$$\tilde{P}(A) = \mathbf{E} X I_A = \lim_{n \to \infty} \mathbf{E} X_n I_A = 0.$$

Definition 5.1 Relation (5.3) between two measures P and $\tilde{P}$ (not necessarily probability measures) is called the absolute continuity of $\tilde{P}$ with respect to P, and denoted $\tilde{P} \ll P$. Moreover, if $P \ll \tilde{P}$ then measures P and $\tilde{P}$ are equivalent ($P \sim \tilde{P}$).

It turns out, one can characterize the absolute continuity property with the help of relations (5.1)-(5.2). This fundamental fact is known as *Radon-Nikodym theorem*.

Theorem 5.1 *Let $(\Omega, \mathcal{F})$ be a measurable space. Let μ be a σ-additive measure and λ be a signed measure which is absolute continuous with respect to μ, $\lambda = \lambda_1 - \lambda_2$, where one of λ_1 or λ_2 is finite. Then there exists $\mathcal{F}$-measurable function $Z = Z(\omega)$, $\omega \in \Omega$, with values in $[-\infty, +\infty]$ such that*

$$\lambda(A) = \int_A Z(\omega) d\mu, \ A \in \mathcal{F}. \qquad (5.4)$$

Moreover, the function Z is uniquely defined up to sets of μ-measure zero.

In representation (5.4) the function Z is called the *Radon-Nikodym density* or *derivative* and denoted $Z = \frac{d\lambda}{d\mu}$.

As the first consequence of Theorem 5.1 we get a convenient *rule of changing of measure in expected values*. Let X be a random variable on a probability space $(\Omega, \mathcal{F}, P)$ and $\tilde{P} \ll P$ with the Radon-Nikodym density Z. Then we formally obtain:

$$\tilde{\mathbf{E}} X = \int_\Omega X d\tilde{P} = \int_\Omega X \frac{d\tilde{P}}{dP} dP = \int_\Omega X Z dP = \mathbf{E} X Z.$$

5.2 Conditional expectations and their properties

Now we are ready to introduce the conditional expected value of random variable X with respect to a σ-algebra $\mathcal{Y} \subseteq \mathcal{F}$. In this case we also call $\mathcal{Y}$ sub-σ-algebra of $\mathcal{F}$.

Definition 5.2 Let X be a non-negative random variable and $\mathcal{Y}$ be a sub-σ-algebra of $\mathcal{F}$. Then a random variable $\mathbf{E}(X|\mathcal{Y})$ is called the conditional expectation of X with respect to $\mathcal{Y}$ if $\mathbf{E}(X|\mathcal{Y})$ is $\mathcal{Y}$-measurable and

$$\mathbf{E} X I_A = \mathbf{E}\left(\mathbf{E}(X|\mathcal{Y}) I_A\right) \tag{5.5}$$

for every $A \in \mathcal{Y}$.

In general case, we decompose $X = X^+ - X^-$ and assume that one of random variables $\mathbf{E}(X^+|\mathcal{Y})$ and $\mathbf{E}(X^-|\mathcal{Y})$ is finite, and define

$$\mathbf{E}(X|\mathcal{Y}) = \mathbf{E}(X^+|\mathcal{Y}) - \mathbf{E}(X^-|\mathcal{Y}). \tag{5.6}$$

Existence of $\mathbf{E}(X|\mathcal{Y})$ in (5.6) follows from the Radon-Nikodym theorem. To prove it we define a (signed) measure $\tilde{P}(A) = \mathbf{E} X I_A$ on a measurable space $(\Omega, \mathcal{Y})$ such that $\tilde{P} \ll P$. According to Theorem 5.1 there exists a unique Radon-Nikodym density which can be denoted here $\mathbf{E}(X|\mathcal{Y})$:

$$\tilde{P}(A) = \int_A \mathbf{E}(X|\mathcal{Y}) dP.$$

Let us note that the above equality coincides with (5.5).

Using definition of $\mathbf{E}(X|\mathcal{Y})$ we can easily define:

Conditional Probability w.r. to $\mathcal{Y}$

$$P(B|\mathcal{Y}) = \mathbf{E}(I_B|\mathcal{Y}), \ B \in \mathcal{F},$$

conditional variance w.r. to $\mathcal{Y}$

$$\mathbf{Var}(X|\mathcal{Y}) = \mathbf{E}\left((X - \mathbf{E}(X|\mathcal{Y}))^2 | \mathcal{Y}\right).$$

It is often the σ-algebra $\mathcal{Y}$ is generated by another random variable Y. Therefore, we can define conditional expected value and conditional probability X w.r. to Y as follows

$$\mathbf{E}(X|Y) = \mathbf{E}(X|\mathcal{Y}), P(B|Y) = P(B|\mathcal{Y}), \ B \in \mathcal{F},$$

where $\mathcal{Y} = \mathcal{Y}^Y$.

Let us count the list of properties of conditional expectations.

1. If $X = c$ (a.s.), then $\mathbf{E}(X|\mathcal{Y}) = c$ (a.s.),
2. If $X \leq Y$ (a.s.), then $\mathbf{E}(X|\mathcal{Y}) \leq \mathbf{E}(Y|\mathcal{Y})$ (a.s.),
3. $|\mathbf{E}(X|\mathcal{Y})| \leq \mathbf{E}(|X| \,|\mathcal{Y})$ (a.s.),

4. For random variables X and Y and constants a and b we have
$$\mathbf{E}(aX + bY|\mathcal{Y}) = a\mathbf{E}(X|\mathcal{Y}) + b\mathbf{E}(Y|\mathcal{Y}) \ (a.s.),$$

5. $\mathbf{E}(X|\mathcal{F}_*) = \mathbf{E}X$ $(a.s.)$,
6. $\mathbf{E}(X|\mathcal{F}) = X$ $(a.s.)$,
7. $\mathbf{E}\mathbf{E}(X|\mathcal{Y}) = \mathbf{E}X$,
8. If sub-σ-algebras $\mathcal{Y}_1 \subseteq \mathcal{Y}_2$, then
$$\mathbf{E}\left(\mathbf{E}(X|\mathcal{Y}_2)|\mathcal{Y}_1\right) = \mathbf{E}(X|\mathcal{Y}_1) \ (a.s.),$$

9. If sub-σ-algebras $\mathcal{Y}_2 \subseteq \mathcal{Y}_1$, then
$$\mathbf{E}\left(\mathbf{E}(X|\mathcal{Y}_2)|\mathcal{Y}_1\right) = \mathbf{E}(X|\mathcal{Y}_2) \ (a.s.),$$

10. We say that two sub-σ-algebras $\mathcal{Y}_1$ and $\mathcal{Y}_2$ are *independent*, if for any $B_1 \in \mathcal{Y}_1$ and $B_2 \in \mathcal{Y}_2$ we have $P(B_1 \cap B_2) = P(B_1)P(B_2)$,
    For a random variable X and a sub-σ-algebra $\mathcal{Y}$ we say that X *does not depend on* $\mathcal{Y}$ if $\mathcal{Y}^X$ and $\mathcal{Y}$ are independent. In this case we have
$$\mathbf{E}(X|\mathcal{Y}) = \mathbf{E}X \ (a.s.),$$

if $\mathbf{E}X$ is well-defined.

11. For random variables X and Y such that Y is $\mathcal{Y}$-measurable, $\mathbf{E}|X| < \infty$, $\mathbf{E}|XY| < \infty$, we have
$$\mathbf{E}(XY|\mathcal{Y}) = Y\mathbf{E}(X|\mathcal{Y}) \ (a.s.)$$

12. For a (generalized) sequence of random variables $(X_n)_{n=1,2,...}$ such that $|X_n| \le Y$, $n = 1, 2, \ldots$, $\mathbf{E}Y < \infty$ and $X_n \to X$, $n \to \infty$, a.s. We have $\mathbf{E}(X_n|\mathcal{Y}) \to \mathbf{E}(X|\mathcal{Y})$, $(a.s.)$ $n \to \infty$,
13. If $X_n \ge Y$, $\mathbf{E}Y > -\infty$ and $X_n \uparrow X$, $n \to \infty$, a.s., then $\mathbf{E}(X_n|\mathcal{Y}) \uparrow \mathbf{E}(X|\mathcal{Y})$, $(a.s.)$ $n \to \infty$,
14. If $X_n \le Y$, $\mathbf{E}Y < \infty$ and $X_n \downarrow X$, $n \to \infty$, a.s., then $\mathbf{E}(X_n|\mathcal{Y}) \downarrow \mathbf{E}(X|\mathcal{Y})$, $(a.s.)$ $n \to \infty$,
15. If $X_n \ge Y$, $\mathbf{E}Y > -\infty$, then
$$\mathbf{E}(\liminf_{n \to \infty} X_n|\mathcal{Y}) \le \liminf_{n \to \infty} \mathbf{E}(X_n|\mathcal{Y}), \ (a.s.)$$

16. For a non-negative random variables $(X_n)_{n=1,2,...}$ we have
$$\mathbf{E}\left(\sum_{n=1}^{\infty} X_n \Big| \mathcal{Y}\right) = \sum_{n=1}^{\infty} \mathbf{E}(X_n|\mathcal{Y}) \ (a.s.)$$

Let us show ways of proof of such properties. In case 2) we have for any $A \in \mathcal{Y}$ that
$$\int_A X dP \le \int_A Y dP.$$

5.2 Conditional expectations and their properties

Therefore, we get from (5.5) that

$$\int_A \mathbf{E}(X|\mathcal{Y})dP \le \int_A \mathbf{E}(Y|\mathcal{Y})dP,$$

which means that $\mathbf{E}(X|\mathcal{Y}) \le \mathbf{E}(Y|\mathcal{Y})$ (a.s.)

In case (8) we have for any $A \in \mathcal{Y}_1 \subseteq \mathcal{Y}_2$ that

$$\int_A \mathbf{E}(X|\mathcal{Y}_1)dP = \int_A XdP = \int_A \mathbf{E}(X|\mathcal{Y}_2) = \int_A \mathbf{E}\left(\mathbf{E}(X|\mathcal{Y}_2)|\mathcal{Y}_1\right)dP,$$

which certifies the statement in (8).

Below we provide some comments and detailed calculations of conditional expectations.

Consider $\mathbf{E}(X|Y)$ for two (integrable) random variables X and Y. According to our definition $\mathbf{E}(X|Y)$ is $\mathcal{Y}^Y$-measurable, and by the representation theorem there exists a Borelian function $\phi(\cdot)$ such that

$$\phi(Y(\omega)) = \mathbf{E}(X|Y)(\omega), \quad \omega \in \Omega. \tag{5.7}$$

For $A \in \mathcal{Y}^Y$ we get from (5.7):

$$\int_A XdP = \int_A \mathbf{E}(X|Y)dP = \int_A \phi(Y)dP. \tag{5.8}$$

Further, taking $A = Y^{-1}(B)$, $B \in \mathcal{B}(R^1)$ we have

$$\int_{Y^{-1}(B)} \phi(Y)dP = \int_B \phi(y)dP_Y,$$

and hence

$$\int_{Y^{-1}(B)} XdP = \int_B \phi(y)dP_Y. \tag{5.9}$$

Having equalities (5.7)-(5.9) we can take $\phi(y) = \mathbf{E}(X|Y = y)$, $y \in R^1$.

To provide more details one can consider a special case when the pair (X, Y) admits a joint density $f_{XY}(x, y)$. In such a case we can put

$$f_{X|Y}(x|y) = \frac{f_{xy}(x, y)}{f_Y(y)}$$

and find that for $C \in \mathcal{B}(R^1)$

$$P(X \in C|Y = y) = \int_C f_{X|Y}(x|y)dx$$

and

$$\mathbf{E}(X|Y = y) = \int_{R^1} x f_{X|Y}(x|y)dx.$$

This is a reason that the function $f_{X|Y}(x|y)$ is called a density of a conditional distribution (conditional density).

In case of discrete random variables X and Y with values $x_1, x_2, \ldots$ and $y_1, y_2, \ldots$ correspondingly, denote $\mathcal{Y} = \sigma(Y) = \mathcal{Y}^Y$ and $(\omega : Y(\omega) = y_i) = D_i$, $i = 1, 2, \ldots$ Sets D_i, $i = 1, 2, \ldots$ are *atoms* for P in the sense that $P(D_i) > 0$ and any subset $A \subseteq D_i$ has a probability zero or its completion to D_i (i.e. $D_i \setminus A$) has a probability zero. We can define

$$\mathbf{E}(X|D_i) = \frac{\mathbf{E}XI_{D_i}}{P(D_i)}, \ i = 1, 2, \ldots$$

Then the calculation of $\mathbf{E}(X|Y)$ is reduced to the claim:

$$\mathbf{E}(X|Y) = \mathbf{E}(X|\mathcal{Y}^Y) = \mathbf{E}(X|D_i) \ (a.s.)$$

on the atom D_i, $i = 1, 2, \ldots$

To connect these definitions and results to the general definition we need to check (5.5). In particular, we have

$$\mathbf{E}\mathbf{E}(X|\mathcal{Y}) = \sum_{i=1}^{\infty} \mathbf{E}(X|D_i) P(D_i)$$

$$= \sum_{i=1}^{\infty} \mathbf{E}(XI_{D_i}) = \sum_{i=1}^{\infty} \sum_{j=1}^{\infty} x_j P(X = x_j, Y = y_i)$$

$$= \sum_{j=1}^{\infty} \sum_{i=1}^{\infty} x_j P(X = x_j, Y = y_i)$$

$$= \sum_{j=1}^{\infty} x_j \sum_{i=1}^{\infty} P(X = x_j, Y = y_i) = \sum_{j=1}^{\infty} x_j P(X = x_j)$$

$$= \mathbf{E}X.$$

Let's take note of that calculations of $\mathbf{E}(X|Y_1, \ldots, Y_n)$ can be given in the same way.

Finally, we demonstrate an important application of $\mathbf{E}(X|Y)$. We will interpret Y as an *observable* variable and X as a non-observable. This is a typical situation in many areas, including option pricing theory, where X is a pay-off of option and Y is a stock price.

How can we estimate X based on observations of Y to provide the optimality of the estimate?

A satisfactory solution of the problem can be given as follows.

Let $\phi = \phi(x)$, $x \in R^1$, be a Borelian function. Taking $\phi(Y)$ we get an estimate for X, and we should choose from a variety of such estimates an *optimal* estimate. As a criteria we can take $\mathbf{E}(X - \phi(Y))^2$ and define the *optimal* estimate $\phi^*(Y)$ as follows

$$\mathbf{E}(X - \phi^*(Y))^2 = \inf_{\phi} \mathbf{E}(X - \phi(Y))^2,$$

where we assume $\mathbf{E}X^2 < \infty$ and $\mathbf{E}\phi^2(Y) < \infty$.

5.2 Conditional expectations and their properties

Theorem 5.2 *The optimal estimate has the following representation* $\phi^*(x) = \mathbf{E}(X|Y = x)$.

Proof Taking $\phi^*(Y) = \mathbf{E}(X|Y)$ we have for any other estimate ϕ:

$$\begin{aligned}
\mathbf{E}(X - \phi(Y))^2 &= \mathbf{E}\left[(X - \phi^*(Y)) + (\phi^*(Y) - \phi(Y))\right]^2 \\
&= \mathbf{E}(X - \phi^*(Y))^2 + 2\mathbf{E}(X - \phi^*(Y))(\phi^*(Y) - \phi(Y)) + \mathbf{E}(\phi^*(Y) - \phi(Y))^2 \\
&= \mathbf{E}(X - \phi^*(Y))^2 + \mathbf{E}(\phi^*(Y) - \phi(Y))^2 + 2\mathbf{E}\left[\mathbf{E}\left((X - \phi^*(Y))(\phi^*(Y) - \phi(Y))\right)|Y\right] \\
&= \mathbf{E}(X - \phi^*(Y))^2 + \mathbf{E}(\phi^*(Y) - \phi(Y))^2 \\
&\geq \mathbf{E}(X - \phi^*(Y))^2.
\end{aligned}$$

Chapter 6
Discrete time stochastic analysis: basic results

Abstract Chapter 6 is completely devoted to a discrete time stochastic analysis. It contains the key notions adapted to discrete time like stochastic basis, filtration, predictability, stopping times, martingales, sub- and supermartingales, local densities of probability measures, discrete stochastic integrals and stochastic exponents. It is stated the Doob decomposition for stochastic sequences, maximal inequalities, and other Doob's theorems. The developed martingale technique is further applied to prove several asymptotical properties for martingales and submartingales (see [1], [7], [8], [10], [15], [40], and [45]).

6.1 Basic notions: stochastic basis, predictability and martingales

We have seen already that a sequence of random variables $X_1, X_2, \ldots, X_n$ can be interpreted as a *n-times repetition* (realization) of the underlying random experiment. One can note that the order of appearance of new information is important. For example, such order and such information are valuable in stock exchange trading. One can also observe that a numeration of X_i may not be directly connected to *real (physical) time*, because an *operation time* is often exploited instead of real time in finance. As we know, to provide an accurate work with random variables we need to assume that these random variables are defined on some probability space $(\Omega, \mathcal{F}, P)$. In this setting, for *each fixed outcome* $\omega \in \Omega$ one can observe a sequence of numbers $X_1(\omega), X_2(\omega), \ldots, X_n(\omega), \ldots$, called a *trajectory* or *sample path*. It may present a behavior of stock prices or indexes over given time interval. So, thinking in this way, we must emphasize the difference between probability theory and stochastic/random processes. It can be roughly explained as follows. If theory of probability studies probabilities of occurrence of random events, including the events connected to random variables, theory of stochastic processes considers probabilities of occurrence of trajectories and families of trajectories of stochastic processes. Such a difference calls for a "supplementary equipment" of given probability space with an *information flow* or *filtration*. It is a non-decreasing family $(\mathcal{F}_n)_{n=0,1,...}$ together with $(\Omega, \mathcal{F}, (\mathcal{F}_n)_{n=0,1,...}, P)$ called a *stochastic basis*. Usually, the main σ-algebra $\mathcal{F}$ is determined by the filtration $(\mathcal{F}_n)_{n=0,1,...}$ in the sense that $\mathcal{F}_\infty = \sigma(\cup_n \mathcal{F}_n) = \mathcal{F}$. We sometimes will count that $\mathcal{F}_0$ is trivial, i.e. $\mathcal{F}_0 = \{\emptyset, \Omega\}$. Filtration creates a new class of random variables, called *stopping times*. A random variable $\tau : \Omega \to \{0, 1, \ldots, n, \ldots, \infty\}$ is called a stopping time if for each $n = 0, 1, \ldots$

$$\{\omega : \tau(\omega) \leq n\} \in \mathcal{F}_n,$$

i.e. each value n is taken based on information until time n without having information from the future time.

We can interpret a stopping time τ as a random time, and therefore we can speak about information $\mathcal{F}_\tau$ before this time τ. It is formally realized as follows:

$$\mathcal{F}_\tau = \{A \in \mathcal{F}_\infty : A \cap \{\omega : \tau(\omega) \leq n\} \in \mathcal{F}_n\},$$

and $\mathcal{F}_\tau$ is a σ-algebra.

If two stopping times τ and σ are connected to each other through inequality $\tau \leq \sigma$ (a.s.), then obviously $\mathcal{F}_\tau \subseteq \mathcal{F}_\sigma$.

Now if $(X_n)_{n=0,1,...}$ is a sequence of random variables on $(\Omega, \mathcal{F}, (\mathcal{F}_n)_{n=0,1,...}, P)$ with information flow $(\mathcal{F}_n)$ and $X_n - \mathcal{F}_n$-measurable for each $n = 0, 1, \ldots$. In this case we will call (X_n) *adapted* to $(\mathcal{F}_n)_{n=0,1,...}$, or *stochastic*.

For a stochastic sequence $(X_n)_{n=0,1,...}$ and a stopping time τ we define

$$X_\tau = \sum_{n=0}^{\infty} X_n I_{\{\tau=n\}} + X_\infty I_{\{\tau=\infty\}},$$

where X_∞ may be a fixed constant or $X_\infty = \lim_{n \to \infty} = X_n$ (a.s.), if the limit does exist.

The random variable X_τ is a "superposition" of a stochastic sequence and τ, and hence, $X_\tau - \mathcal{F}_\tau$-measurable. We can go even further if we take a stochastic sequence $(X_n)_{n=0,1,...}$ and a non-decreasing sequence of stopping times $(\tau_n)_{n=0,1,...}$ and define a new sequence of random variables

$$Y_n = X_{\tau_n}, \quad \tau_n \leq \tau_{n+1}.$$

It is quite natural to call (τ_n) as a *time change* and (Y_n) as a time changed sequence. The idea of "time change" is very productive for Stochastic Analysis. For example, in Mathematical Finance this type of time change is often based on market volatility, and is called the "operation time".

Problem 6.1 Prove that
1. τ is a stopping time if and only if $\{\tau = n\} \in \mathcal{F}_n$ for all n.
2. $\mathcal{F}_\tau$ is a σ-algebra, if τ is a stopping time.
3. $\mathcal{F}_\tau \subseteq \mathcal{F}_\sigma$ if stopping times $\tau \leq \sigma(\omega)$ (a.s.).
4. X_τ is $\mathcal{F}_\tau$-measurable if τ is a stopping time.
5. $\tau \wedge \sigma$ is a stopping time.

6.1 Basic notions: stochastic basis, predictability and martingales

We can also consider $(\Omega, \mathcal{F}, (\mathcal{F}_n), P)$ as a system of probability spaces $(\Omega, \mathcal{F}_0, P_0), (\Omega, \mathcal{F}_1, P_1), \ldots$, where P_0 on $\mathcal{F}_0$ is just a restriction of P to $\mathcal{F}_0$ and so on. In this framework we can consider another probability measure $\tilde{P}$ and $\tilde{P}_n$ as its restriction to $\mathcal{F}_n$. Assume that $\tilde{P}_n \ll P_n$, $n = 0, 1, \ldots$ and denote $Z_n = \frac{d\tilde{P}_n}{dP_n}$ the density of $\tilde{P}_n$ with respect to P_n, $n = 0, 1, \ldots$. The sequence $(Z_n)_{n=0,1,\ldots}$ is called a *local density* of $\tilde{P}$ with respect to P, and for each $n = 0, 1, \ldots$ we have

$$\mathbf{E} Z_n = \int_\Omega Z_n \, dP = \int_\Omega Z_n \, dP_n = \tilde{P}_n(\Omega) = 1.$$

Further, for any $A \in \mathcal{F}_{n-1}$ we formally have that

$$\int_A Z_n \, dP = \int_A Z_n \, dP_n = \int_A \frac{d\tilde{P}_n}{dP_n} dP_n = \tilde{P}_n(A) = \tilde{P}_{n-1}(A) = \int_A Z_{n-1} \, dP.$$

It means that (a.s.)

$$\mathbf{E}(Z_n | \mathcal{F}_{n-1}) = Z_{n-1}, \quad n = 1, 2, \ldots \tag{6.1}$$

We can also ask the question:

Is it possible to calculate for some integrable random variable Y the conditional expectation $\tilde{\mathbf{E}}(Y|\mathcal{F}_{n-1})$ via $\mathbf{E}(Y|\mathcal{F}_{n-1})$?

Again, take $A \in \mathcal{F}_{n-1}$ and find that

$$\int_A \mathbf{E}(YZ_n|\mathcal{F}_{n-1}) dP = \int_A Y Z_n \, dP = \int_A Y \, d\tilde{P}_n$$

$$= \int_A Y \, d\tilde{P}_{n-1} = \int_A \tilde{\mathbf{E}}(Y|\mathcal{F}_{n-1}) d\tilde{P}_{n-1}$$

$$= \int_A \tilde{\mathbf{E}}(Y|\mathcal{F}_{n-1}) dP_{n-1} = \int_A \tilde{\mathbf{E}}(Y|\mathcal{F}_{n-1}) Z_{n-1} \, dP,$$

and hence,

$$\tilde{\mathbf{E}}(Y|\mathcal{F}_{n-1}) = Z_{n-1}^{-1} \mathbf{E}(YZ_n|\mathcal{F}_{n-1}) \quad (a.s.) \tag{6.2}$$

Relation (6.2) is called a *rule of change of probability in conditional expectations*.

Besides adapted sequences of random variables, we can introduce sequences of random variables which are in between deterministic and stochastic sequences. We say that an adapted sequence $(A_n)_{n=0,1,\ldots}$ is *predictable*, if A_n-$\mathcal{F}_{n-1}$-measurable for $n = 1, 2, \ldots$. We also can take the property (6.1) to introduce the whole class of stochastic sequences called *martingales*. We say that an integrable adapted sequence $(M_n)_{n=0,1,\ldots}$ is a martingale, if

$$\mathbf{E}(M_n|\mathcal{F}_{n-1}) = M_{n-1} \quad (a.s.) \quad for \ n = 1, 2, \ldots. \tag{6.3}$$

Relation (6.3) can be rewritten as follows $\mathbf{E}(M_n - M_{n-1}|\mathcal{F}_{n-1}) = 0$ $(a.s.)$ It means the sequence $Y_n = M_n - M_{n-1}$ presents so-called a *martingale-difference*.

We can see also that the local density $(Z_n)_{n=0,1,\ldots}$ in (6.1) is a martingale with respect to P. These two types of stochastic sequences is a natural basis for many others. To demonstrate this claim, we consider an arbitrary integrable stochastic sequence $(X_n)_{n=0,1,\ldots}$ with $X_0 = 0$ (a.s.) for simplicity. Define for each $n = 1, 2, \ldots$, $\Delta X_n = X_n - X_{n-1}$ and write an obvious equality

$$\Delta X_n = \Delta X_n - \mathbf{E}(\Delta X_n | \mathcal{F}_{n-1}) + \mathbf{E}(\Delta X_n | \mathcal{F}_{n-1}) = \Delta M_n + \Delta A_n; \tag{6.4}$$

where

$$\Delta M_n = \Delta X_n - \mathbf{E}(\Delta X_n | \mathcal{F}_{n-1}),$$

$$\Delta A_n = \mathbf{E}(\Delta X_n | \mathcal{F}_{n-1}).$$

It is clear from (6.4) that for $n = 0, 1, \ldots$,

$$X_n = M_n + A_n, \tag{6.5}$$

where $M_n = \sum_{i \leq n} \Delta M_i$ and $A_n = \sum_{i \leq n} \Delta A_i$; and therefore, $(M_n)_{n=0,1,\ldots}$ is a martingale and $(A_n)_{n=0,1,\ldots}$ is predictable.

Such a unique decomposition (6.5) is called the *Doob decomposition* of $(X_n)_{n=0,1,\ldots}$ In particular, it is true for sequences $(X_n)_{n=0,1,\ldots}$ satisfying conditions

$$\mathbf{E}(X_n | \mathcal{F}_{n-1}) \geq X_{n-1} \quad (a.s.)$$

and

$$\mathbf{E}(X_n | \mathcal{F}_{n-1}) \leq X_{n-1} \quad (a.s.)$$

Such sequences are called *submartingales* and *supermartingales* and in their decompositions (6.5) $\Delta A_n \geq 0$ and $\Delta A_n \leq 0$ (a.s.), $n = 1, 2, \ldots$, correspondingly.

Remark 6.1 Given a martingale $(X_n)_{n=0,1,\ldots}$ there is a simple way of constructing submartingales (supermartingales). Suppose ϕ is a convex downward measurable function such that $\mathbf{E}|\phi(X_n)| < \infty$, $n = 0, 1, \ldots$ Then Jensen's inequality implies that $(\phi(X_n))_{n=0,1,\ldots}$ is a submartingale.

Let us consider a martingale $(M_n)_{n=0,1,\ldots}$ such that $\mathbf{E}M_n^2 < \infty$, $n = 0, 1, \ldots$, then it is called *square-integrable*. Further, due to the Jensen's inequality $X_n = M_n^2$ is a submartingale. Using the Doob decomposition (6.5) one can conclude that

$$M_n^2 = m_n + \langle M, M \rangle_n, \quad n = 0, 1, \ldots,$$

where $(m_n)_{n=0,1,\ldots}$ is a martingale and $(\langle M, M \rangle_n)_{n=0,1,\ldots}$ is a non-decreasing predictable sequence called the *quadratic characteristic (compensator)* of M. Moreover,

$$\langle M, M \rangle_n = \sum_{i=1}^{n} \mathbf{E}((\Delta M_i)^2 | \mathcal{F}_{i-1}), \quad \langle M, M \rangle_0 = 0,$$

6.1 Basic notions: stochastic basis, predictability and martingales

and

$$\mathbf{E}\left((M_k - M_l)^2|\mathcal{F}_l\right) = \mathbf{E}(M_k^2 - M_l^2|\mathcal{F}_l)$$
$$= \mathbf{E}(\langle M,M \rangle_k - \langle M,M \rangle_l|\mathcal{F}_l), \ l \le k,$$

and $\mathbf{E}M_n^2 = \mathbf{E}\langle M, M\rangle_n$, $n = 0, 1, \ldots$. It is possible to define a measure of correlation between two square-integrable martingales $(M_n)_{n=0,1,\ldots}$ and $(N_n)_{n=0,1,\ldots}$:

$$\langle M, N \rangle_n = \frac{1}{4}\{\langle M+N, M+N \rangle_n - \langle M-N, M-N \rangle_n\},$$

$n = 0, 1, \ldots$, which is called the *joint quadratic characteristic* of $(M_n)_{n=0,1,\ldots}$ and $(N_n)_{n=0,1,\ldots}$.

It is almost obvious to show that the sequence $(M_n N_n - \langle M, N\rangle_n)_{n=0,1,\ldots}$ is a martingale, and if $\langle M, N \rangle_n = 0$ for all $n = 0, 1, \ldots$, then such martingales are called *orthogonal*.

Further, the following property that connects the martingale property and absolute continuity of probability measures is related to formula (6.2) and can be referred to as a discrete time version of the *Girsanov theorem*.

Let $(M_n)_{n=0,1,\ldots}$, $M_0 = 0$, be a martingale with respect to the original measure P and assume $\mathbf{E}|Z_n Z_{n-1}^{-1} \Delta M_n| < \infty$, $n = 1, 2, \ldots$. Define $(\tilde{M}_n)_{n=0,1,\ldots}$, $\tilde{M}_0 = 0$, by relations

$$\Delta \tilde{M}_n = \Delta M_n - \mathbf{E}(Z_n Z_{n-1}^{-1} \Delta M_n | \mathcal{F}_{n-1}), \ n = 1, 2, \ldots.$$

Using the rule of change of measure (6.2), we obtain

$$\tilde{\mathbf{E}}(\Delta \tilde{M}_n | \mathcal{F}_{n-1}) = \tilde{\mathbf{E}}(\Delta M_n - \mathbf{E}(Z_n Z_{n-1}^{-1} \Delta M_n | \mathcal{F}_{n-1}) | \mathcal{F}_{n-1})$$
$$= \tilde{\mathbf{E}}(\Delta M_n | \mathcal{F}_{n-1}) - \tilde{\mathbf{E}}(\Delta M_n | \mathcal{F}_{n-1}) = 0 \ (a.s.),$$

which implies that $(\tilde{M}_n)_{n=0,1,\ldots}$ is a martingale with respect to $\tilde{P} \ll P$ with the local density $(Z_n)_{n=0,1,\ldots}$.

Let us use predictable and martingale stochastic sequences to construct more complicated objects. For a predictable $(H_n)_{n=0,1,\ldots}$ and a martingale $(m_n)_{n=0,1,\ldots}$ define a *discrete stochastic integral*

$$H * m_n = \sum_{i=0}^{n} H_i \Delta m_i. \tag{6.6}$$

If martingale $(m_n)_{n=0,1,\ldots}$ is square-integrable, sequence $(H_n)_{n=0,1,\ldots}$ is predictable and $\mathbf{E}H_n^2 \Delta \langle m, m, \rangle_n < \infty$, $n = 0, 1, \ldots$, then $(H * M_n)_{n=0,1,\ldots}$ is a square-integrable martingale with quadratic characteristic

$$\langle H*m, H*m \rangle_n = \sum_{i=0}^{n} H_i^2 \Delta \langle m, m \rangle_i.$$

Further, let $(M_n)_{n=0,1,...}$ be a fixed square-integrable martingale, then one can consider all square-integrable martingales $(N_n)_{n=0,1,...}$ that are orthogonal to $(M_n)_{n=0,1,...}$. Introduce a family of square-integrable martingales of the form

$$X_n = M_n + N_n. \tag{6.7}$$

On the other hand, any square-integrable martingale $(X_n)_{n=0,1,...}$ can be written in the form (6.7), where the orthogonal term $(N_n)_{n=0,1,...}$ satisfies (6.6) with the martingale $(m_n)_{n=0,1,...}$ that is orthogonal to the given martingale $(M_n)_{n=0,1,...}$. Such a version of decomposition (6.7) is referred to as the *Kunita-Watanabe decomposition*.

Discrete stochastic integrals are related to *discrete stochastic differential equations*. Solutions of such equations are used in modeling the dynamics of asset prices in financial markets.

Consider a stochastic sequence $(U_n)_{n=0,1,...}$ with $U_0 = 0$ and define a new stochastic sequence $(X_n)_{n=0,1,...}$ with $X_0 = 1$ by

$$\Delta X_n = X_{n-1}\Delta U_n, \ n = 1, 2, \ldots. \tag{6.8}$$

Solution of (6.8) has the form

$$X_n = \prod_{i=1}^{n}(1 + \Delta U_i) = \mathcal{E}_n(U), \ n = 1, 2, \ldots,$$

and is called a *stochastic exponential*.

One can consider a non-homogeneous version of equation (6.8)

$$\Delta X_n = \Delta N_n + X_{n-1}\Delta U_n, \ X_0 = N_0, \tag{6.9}$$

where $(N_n)_{n=0,1,...}$ is a given sequence.

Solution of (6.9) is determined with the help of stochastic exponential as follows

$$X_n = \mathcal{E}_n(U)\left[N_0 + \sum_{i=1}^{n}\mathcal{E}_i^{-1}(U)\Delta U_i\right].$$

Let us list very helpful properties of stochastic exponentials.

1. $\mathcal{E}_n^{-1}(U) = \mathcal{E}_n(-U^*)$, where $\Delta U_n^* = \frac{\Delta U_n}{1+\Delta U_n}$, $\Delta U_n \neq -1$;
2. $(\mathcal{E}_n(U))_{n=0,1,...}$ is a martingale if and only if $(U_n)_{n=0,1,...}$ is a martingale;
3. $\mathcal{E}_n(U) = 0$ (a.s.) for $n \geq \tau_0 = \inf(i : \mathcal{E}_i(U) = 0)$;
4. For two stochastic sequences (U_n) and (V_n) the next multiplication rule of stochastic exponentials is true:

$$\mathcal{E}_n(U)\mathcal{E}_n(V) = \mathcal{E}_n(U + V + [U, V]),$$

where $\Delta[U, V]_n = \Delta U_n \Delta V_n$.

6.2 Martingales on finite time interval

Here we study stochastic sequences on the interval $[0, N] = \{0, 1, \ldots, N\}$. The first important theorem is devoted to an interesting characterization of the notion of a martingale.

Theorem 6.1 *Let $(X_n)_{n=0,1,\ldots,N}$ be an integrable stochastic sequence on a stochastic basis $(\Omega, \mathcal{F}, (\mathcal{F}_n)_{n=0,1,\ldots,N}, P)$. Then the following statements are true*

1) *The sequence $(X_n)_{n=0,1,\ldots,N}$ is a martingale if and only if $X_n = \mathbf{E}(X_N|\mathcal{F}_n)$ (a.s.) for all $n = 0, 1, \ldots, N$.*
2) *If for all stopping times τ the equality $\mathbf{E}X_\tau = \mathbf{E}X_0$ is fulfilled then $(X_n)_{n=0,1,\ldots,N}$ is a martingale.*

Proof 1) For a direct implication, using the definition of conditional expectations and their telescopic property, we have that (a.s.)

$$\mathbf{E}(X_N|\mathcal{F}_{N-2}) = \mathbf{E}\left(\mathbf{E}(X_N|\mathcal{F}_{N-1})|\mathcal{F}_{N-2}\right) = \mathbf{E}(X_{N-1}|\mathcal{F}_{N-2}) = X_{N-2}, \text{ etc.}$$

For an inverse implication we also use the definition and the telescopic property and find that for all n (a.s.)

$$\mathbf{E}(X_N|\mathcal{F}_{N-1}) = \mathbf{E}\left(\mathbf{E}(X_N|\mathcal{F}_N)|\mathcal{F}_{N-1}\right) = \mathbf{E}(X_N|\mathcal{F}_{N-1}) = X_{N-1}, \text{ etc.}$$

2) For a fixed $n \leq N$ and $A \in \mathcal{F}_n$ we define the following stopping time

$$\tau_A(\omega) = \begin{cases} n & if\ \omega \in A, \\ N & if\ \omega \notin A. \end{cases}$$

According to the assumption we have

$$\mathbf{E}X_0 = \mathbf{E}X_{\tau_A} = \mathbf{E}X_{\tau_A}I_A + \mathbf{E}X_{\tau_A}I_{A^c}$$
$$= \mathbf{E}X_n I_A + \mathbf{E}X_N I_{A^c}.$$

It means that $\mathbf{E}(X_N|\mathcal{F}_n) = X_n$ (a.s.) and by (a) we conclude that $(X_n)_{n=0,1,\ldots,N}$ is a martingale. □

Theorem 6.1 gives us an understanding about a strong connection between martingales and stopping times. The next statement which has a special name as "optional sampling theorem" of Doob tells us more about these connections.

Theorem 6.2 *Let $(X_n)_{n=0,1,\ldots,N}$ be a martingale (submartingale, supermartingale) and $\tau_1 \leq \tau_2$ (a.s.) are stopping times, then*

$$\mathbf{E}(X_{\tau_2}|\mathcal{F}_{\tau_1}) = X_{\tau_1}\ (a.s.)$$

($\mathbf{E}(X_{\tau_2}|\mathcal{F}_{\tau_1}) \geq X_{\tau_1}$ and $\mathbf{E}(X_{\tau_2}|\mathcal{F}_{\tau_1}) \leq X_{\tau_1}$ (a.s.), respectively). In particular $\mathbf{E}X_{\tau_1} = \mathbf{E}X_{\tau_2}$.

Proof of this theorem, which generalizes the obvious property for deterministic times, readily follows from next considerations.

For $A \in \mathcal{F}_{\tau_1}$, $n \leq N$ and $B = A \cap \{\tau_1 = n\}$ in order to prove $\int_A X_{\tau_2} dP = \int_A X_{\tau_1} dP$ we need to prove that

$$\int_{B \cap \{\tau_2 \geq n\}} X_{\tau_2} dP = \int_{B \cap \{\tau_2 \geq n\}} X_n dP.$$

It follows from the next equalities

$$\int_{B \cap \{\tau_2 \geq n\}} X_n dP = \int_{B \cap \{\tau_2 = n\}} X_n dP + \int_{B \cap \{\tau_2 > n\}} X_n dP$$

$$= \int_{B \cap \{\tau_2 = n\}} X_n dP + \int_{B \cap \{\tau_2 > n\}} \mathbf{E}(X_{n+1}|\mathcal{F}_n) dP$$

$$= \int_{B \cap \{\tau_2 = n\}} X_n dP + \int_{B \cap \{\tau_2 > n\}} X_{n+1} dP$$

$$= \int_{B \cap \{n \leq \tau_2 \leq n+1\}} X_n dP + \int_{B \cap \{\tau_2 \geq n+2\}} X_{n+2} dP = \ldots =$$

$$= \int_{B \cap \{\tau_2 \geq n\}} X_{\tau_2} dP.$$

It turns out for martingales (submartingales, supermartingales) one can get inequalities called *maximal* (or the *Kolmogorov-Doob inequalities*) which are stronger than the Chebyshev inequality.

Theorem 6.3 *1. If $(X_n)_{n=0,1,\ldots,N}$ is a submartingale, then for any $\lambda > 0$:*

$$P(\omega : \max_{n \leq N} X_n \geq \lambda) \leq \frac{\mathbf{E}X_N^+}{\lambda}.$$

2. If $(X_n)_{n=0,1,\ldots,N}$ is a supermartingale, then for any $\lambda > 0$:

$$P(\omega : \max_{n \leq N} X_n \geq \lambda) \leq \frac{\mathbf{E}X_0 + \mathbf{E}X_N^-}{\lambda}.$$

3. If $(X_n)_{n=0,1,\ldots,N}$ is a martingale, then for any $\lambda > 0$:

$$P(\omega : \max_{n \leq N} |X_n| \geq \lambda) \leq \frac{\mathbf{E}|X_N|}{\lambda}.$$

Proof Let us define the following stopping time

$$\tau = \inf\{n \leq N : X_n \geq \lambda\},$$

6.2 Martingales on finite time interval

where we put $\tau = N$ if the set in brackets above is $\emptyset$.

Denote $A = \{\omega : \max_{n \leq N} X_n \geq \lambda\}$ and find that

$$A \cap \{\tau = n\} = \{\tau = n\} \in \mathcal{F}_n,$$

if $n < N$ and

$$A \cap \{\tau = N\} \in \mathcal{F}_N,$$

due to $A \in \mathcal{F}_N$.

Hence, $A \in \mathcal{F}_\tau$.

To prove the statement (1) of the theorem we derive from Theorem 6.2 that

$$\mathbf{E}X_\tau I_A \leq \mathbf{E}X_N I_A$$

because stopping time $\tau \leq N$.

If the event A occurs, then $X_\tau \geq \lambda$ and $X_\tau I_A \geq \lambda I_A$. Therefore,

$$\lambda P(A) \leq \mathbf{E}X_N I_A \leq \mathbf{E}X_N^+ I_A \leq \mathbf{E}X_N^+,$$

and we obtain (1).

Further, due to $0 \leq \tau \leq N$ we have from Theorem 6.2

$$\mathbf{E}X_0 \geq \mathbf{E}X_\tau = \mathbf{E}X_\tau I_A + \mathbf{E}X_\tau I_{A^c} \geq \mathbf{E}X_\tau I_A + \mathbf{E}X_N I_{A^c}.$$

Hence, we get statement (2) after next calculations

$$\lambda P(A) \leq \mathbf{E}X_\tau I_A \leq \mathbf{E}X_0 - \mathbf{E}X_N I_{A^c}$$
$$= \mathbf{E}X_0 + \mathbf{E}(-X_N) I_{A^c} \leq \mathbf{E}X_0 + \mathbf{E}(-X_N)^+ I_{A^c}$$
$$\leq \mathbf{E}X^0 + \mathbf{E}X_N^-.$$

The last statement (3) is a combination of (1) and (2). □

We finish this section by the well-known and helpful results of Doob about estimation of number of upcrossings (downcrossings) of given interval by submartingales and supermartingales.

Let $a < b$ be real numbers and $(X_n)_{n=0,1,\ldots,N}$ be a stochastic sequence. Define the following sequence of stopping times:

$$\tau_0 = 0, \quad \tau_1 = \inf\{n > 0 : X_n \leq a\},$$
$$\tau_2 = \inf\{n > \tau_1 : X_n \geq b\}, \ldots,$$
$$\tau_{2m-1} = \inf\{n > \tau_{2m-2} : X_n \leq a\},$$
$$\tau_{2m} = \inf\{n > \tau_{2m-1} : X_n \geq b\},$$

where $\tau_k = N$ if the set in brackets above is $\emptyset$.

It is clear that between τ_{2m-1} and τ_{2m} that is an *upcrossing* of the interval (a, b) by the sequence (X_n) occurs.

Define the following random variable

$$\beta_+^X(N, a, b) = \begin{cases} 0, & if\ \tau_2 > N, \\ \max\{m : \tau_{2m} \le N\}, & if\ \tau_2 \le N, \end{cases}$$

which is called the *number of upcrossings* of (a, b) by the sequence (X_n) during the time interval $[0, N]$.

Definition of notion of downcrossings is given in a similar way. Denote the corresponding number of downcrossings by $\beta_-^X(N, a, b)$.

Theorem 6.4 *Let* $(X_n)_{n=0,1,\ldots,N}$ *be a submartingale and* $\beta_\pm^X(N, a, b)$ *be the number of upcrossings (downcrossings) of the interval* (a, b). *Then*

$$E\beta_+^X(N, a, b) \le \frac{E(X_N - a)^+}{b - a}, \tag{6.10}$$

$$E\beta_-^X(N, a, b) \le \frac{E(X_N - b)^+}{b - a}. \tag{6.11}$$

Proof We prove only (6.10) in view a symmetry of formulas (6.10) and (6.11). Let us reduce the initial problem to estimation of number of upcrossings by a non-negative submartingale $((X_n - a)^+)_{n=0,1,\ldots,N}$ of the interval $(0, b - a)$. Moreover, we can put $a = 0$ and prove that

$$E\beta_+^X(N, 0, b) \le \frac{EX_N}{b}, \tag{6.12}$$

for a non-negative submartingale $(X_n)_{n=0,1,2,\ldots,N}$, $X_0 = 0$.

Define for $i = 1, 2, \ldots$ a sequence of random variables:

$$\phi_i = \begin{cases} 1, & if\ \tau_m < i \le \tau_{m+1}\ \text{for some non-even } m, \\ 0, & if\ \tau_m < i \le \tau_{m+1}\ \text{for some even } m. \end{cases}$$

It follows from definitions of $\beta_+^X(N, a, b)$ and ϕ_i that

$$b\beta_+^X(N, 0, b) \le \sum_{i=1}^N \phi_i(X_i - X_{i-1}).$$

Let us note that

$$\{\phi_i = 1\} = \cup_{non-even\ m}(\{\tau_m < i\} \notin \{\tau_{m+1} < i\}) \in \mathcal{F}_{i-1},$$

and therefore by properties of conditional expected values and submartingales

$$bE\beta_+^X(N,0,b) \leq E\sum_{i=1}^{N}\phi_i(X_i - X_{i-1})$$

$$= \sum_{i=1}^{N}\int_{\{\phi_i\}}(X_i - X_{i-1})dP$$

$$= \sum_{i=1}^{N}\int_{\{\phi_i\}}\mathbf{E}(X_i - X_{i-1}|\mathcal{F}_{i-1})dP$$

$$\leq \sum_{i=1}^{N}\mathbf{EE}(X_i - X_{i-1}|\mathcal{F}_{i-1})$$

$$= (EX_1 - EX_0) + (EX_2 - EX_1) + \ldots + (EX_N - EX_{N-1}) = EX_N.$$

and we arrive to (6.12), and further to (6.10). □

Corollary 6.1 *Let $(X_n)_{n=0,1,\ldots,N}$ be a supermartingale and $\alpha_+^X(N,a,b)$ be the number of upcrossings of the interval (a,b). Then*

$$E\alpha_+^X(N,a,b) \leq \frac{E(a - X_N)^+}{b - a}. \tag{6.13}$$

Proof The inequality (6.13) follows from (6.11) because $\alpha_+^X(N,a,b)$ can be interpreted as the number of downcrossings of the interval $(-b,-a)$ by the submartingale $(-X_n)_{n=0,1,\ldots,N}$:

$$E\alpha_+^X(N,a,b) = E\beta_-^{-X}(N,-b,-a) \leq \frac{E(-X_N - (-a))^+}{(-a) - (-b)} = \frac{E(a - X_N)^+}{b - a}.$$

□

6.3 Martingales on infinite time interval

Here martingales (submartingales, supermartingales) are studied for $\mathbb{Z}_+ = \{0, 1, \ldots, n, \ldots\}$. Statements of Theorem 6.3 and Theorem 6.4 are transformed by taking the limits as $N \to \infty$, and we get the following list of inequalities (6.14)-(6.19):

$$P(\omega : \sup_n X_n \geq \lambda) \leq \frac{\sup_n EX_n^+}{\lambda}, \quad \lambda > 0, \tag{6.14}$$

for a submartingale $(X_n)_{n=0,1,\ldots}$ with $\sup_n EX_n^+ < \infty$;

$$P(\omega : \sup_n X_n \geq \lambda) \leq \frac{EX_0 + \sup_n EX_n^-}{\lambda}, \quad \lambda > 0, \tag{6.15}$$

for a supermartingale $(X_n)_{n=0,1,...}$ with $\sup_n \mathbf{E} X_n^- < \infty$;

$$P(\omega : \sup_n |X_n| \geq \lambda) \leq \frac{\sup_n \mathbf{E}|X|_n}{\lambda}, \quad \lambda > 0, \qquad (6.16)$$

for a martingale $(X_n)_{n=0,1,...,N}$ with $\sup_n \mathbf{E}|X|_n < \infty$;

$$\mathbf{E}\beta_+^X(\infty, a, b) \leq \frac{\sup_n \mathbf{E}(X_n - a)^+}{b - a}, \qquad (6.17)$$

for a submartingale $(X_n)_{n=0,1,...}$ with $\sup_n \mathbf{E}(X_n - a)^+ < \infty$;

$$\mathbf{E}\beta_-^X(\infty, a, b) \leq \frac{\sup_n \mathbf{E}(X_n - b)^+}{b - a}. \qquad (6.18)$$

for a submartingale $(X_n)_{n=0,1,...}$ with $\sup_n \mathbf{E}(X_n - b)^+ < \infty$;

$$\mathbf{E}\alpha_+^X(\infty, a, b) \leq \frac{\sup_n \mathbf{E}(a - X_n)^+}{b - a}. \qquad (6.19)$$

for a supermartingale $(X_n)_{n=0,1,...}$ with $\sup_n \mathbf{E}(a - X_n)^+ < \infty$.

The *Doob optional stopping (sampling) theorem* admits a natural extension to $\mathbb{Z}_+$ too.

Theorem 6.5 *Let $(X_n)_{n=0,1,...}$ be a uniformly integrable martingale (submartingale, supermartingale) and τ be a finite stopping time. Then $\mathbf{E}|X_\tau| < \infty$ and*

$$\mathbf{E} X_\tau = \mathbf{E} X_0 \ (\mathbf{E} X_\tau \geq \mathbf{E} X_0, \mathbf{E} X_\tau \leq \mathbf{E} X_0). \qquad (6.20)$$

Moreover, for finite stopping times $\tau \leq \sigma$ (a.s.):

$$\mathbf{E}(X_\sigma | \mathcal{F}_\tau) = X_\tau \ (a.s.), \ (\mathbf{E}(X_\sigma | \mathcal{F}_\tau) \geq X_\tau \ (a.s.), \ \mathbf{E}(X_\sigma | \mathcal{F}_\tau) \leq X_\tau \ (a.s.)). \quad (6.21)$$

Proof Both formulas (6.20) and (6.21) are proved with the help of limit arguments. Let us only provide the proof of (6.20).

For a fixed N denote $\tau_N = \tau \wedge N$ which is a bounded stopping time, and by Theorem 6.2 we have

$$\mathbf{E} X_0 = \mathbf{E} X_{\tau_N}. \qquad (6.22)$$

Let us note from (6.22) that

$$\mathbf{E} X_{\tau_N} = 2\mathbf{E} X_{\tau_N}^+ - \mathbf{E} X_{\tau_N} \leq 2\mathbf{E} X_{\tau_N}^+ - \mathbf{E} X_0. \qquad (6.23)$$

Further, using (6.23) and a submartingale property of $(X_n^+)_{n=0,1,...}$ we obtain that

6.3 Martingales on infinite time interval

$$\mathbf{E}X^+_{\tau_N} = \sum_{j=0}^{N} \mathbf{E}X^+_j I_{\{\tau_N=j\}} + \int_{\{\tau>N\}} X^+_N dP$$

$$\leq \sum_{j=0}^{N} \mathbf{E}X^+_j I_{\{\tau_N=j\}} + \mathbf{E}X^+_N I_{\{\tau>N\}}$$

$$= \mathbf{E}X^+_N \leq \mathbf{E}|X_N| \leq \sup_n \mathbf{E}|X_n|,$$

and, hence,

$$\mathbf{E}|X_{\tau_N}| \leq 3 \sup_n \mathbf{E}|X_n| < \infty. \qquad (6.24)$$

Using (6.24) and the Fatou lemma we get

$$\mathbf{E}|X_\tau| \leq \limsup_N \mathbf{E}|X_{\tau_N}| \leq 4 \sup_n \mathbf{E}|X_n| < \infty.$$

Let us note that for a finite stopping time τ we have

$$\lim_{n\to\infty} \mathbf{E}|X_n| I_{\{\tau>n\}} = 0, \qquad (6.25)$$

because $(X_n)_{n=0,1,\ldots}$ is uniformly integrable and $P(\omega : \tau(\omega) > n) \to 0$, $n \to \infty$. We can write a decomposition

$$X_\tau = X_{\tau \wedge n} + (X_\tau - X_n)I_{\{\tau>n\}},$$

and its average

$$\mathbf{E}X_\tau = \mathbf{E}X_{\tau \wedge n} + \mathbf{E}X_\tau I_{\{\tau>n\}} - \mathbf{E}X_n I_{\{\tau>n\}}, \qquad (6.26)$$

In the equation (6.26) $\mathbf{E}X_{\tau \wedge n} = \mathbf{E}X_0$, because $(X_{\tau \wedge n})$ is a martingale.
The second term of (6.26)

$$\mathbf{E}(X_\tau)I_{\{\tau>n\}} = \sum_{i=n+1}^{\infty} \mathbf{E}X_i I_{\{\tau=i\}} \to 0, n \to \infty,$$

because the series

$$\mathbf{E}X_\tau = \sum_{i=0}^{\infty} \mathbf{E}X_i I_{\{\tau=i\}} \text{ converges.}$$

The third term of (6.26) converges to zero because (6.25). □

Let us prove a key theorem of Doob about (a.s.)-convergence of submartingales.

Theorem 6.6 *Let $(X_n)_{n=0,1,\ldots}$ be a submartingale satisfying condition*

$$\sup_n \mathbf{E}|X_n| < \infty. \qquad (6.27)$$

Then there exists an integrable random variable $X_\infty = \lim_{n\to\infty} X_n$ (a.s.).

Proof Assume that this limit does not exist. It means that
$$P(\omega : \limsup_{n\to\infty} X_n(\omega) > \liminf_{n\to\infty} X_n(\omega)) > 0. \tag{6.28}$$

The set in (6.28) can be written
$$\{\omega : \limsup_n X_n > \liminf_n X_n\} = \cup_{a<b, a,b\in\mathbb{Q}}\{\omega : \limsup_n X_n > b > a > \liminf_n X_n\}, \tag{6.29}$$
where $\mathbb{Q} \subseteq \mathbb{R}^1$ is the set of rational numbers.

It follows from (6.28)-(6.29) that for some rational numbers $a < b$ we have
$$P(\omega : \limsup_n X_n > b > a > \liminf_n X_n) > 0. \tag{6.30}$$

We also note that for each $n \in \mathbb{Z}$
$$\mathbf{E} X_n^+ \leq \mathbf{E}|X_n| = 2\mathbf{E} X_n^+ - \mathbf{E} X_n \leq 2\mathbf{E} X_n^+ - \mathbf{E} X_0,$$

and find that for submartingale $(X_n)_{n\in\mathbb{Z}}$ the following conditions are equivalent:
$$\sup_n \mathbf{E}|X_n| < \infty \Leftrightarrow \sup_n \mathbf{E} X_n^+ < \infty. \tag{6.31}$$

Applying (6.17) together with (6.31) we obtain
$$\mathbf{E}\beta_+^X(\infty, a, b) \leq \frac{\sup_n \mathbf{E} X_n^+ + |a|}{b-a} < \infty. \tag{6.32}$$

Condition (6.32) is a contradiction with the assumption (6.30) and (6.28). Hence, there exists $X_\infty = \lim_{n\to\infty} X_n$ (a.s.) which is integrable by the Fatou lemma and $\mathbf{E}|X_\infty| \leq \sup_n \mathbf{E}|X_n| < \infty$. □

Corollary 6.2 *(a) If $(X_n)_{n=0,1,...}$ is a non-negative martingale, then $\sup_n \mathbf{E}|X_n| = \sup_n \mathbf{E} X_n = \mathbf{E} X_0 < \infty$ and therefore there exists integrable $X_\infty = \lim_{n\to\infty} X_n$ (a.s.).*

(b) If $(X_n)_{n=0,1,...}$ is a non-positive submartingale, then there exists integrable $X_\infty = \lim_{n\to\infty} X_n$ (a.s.) and $(X_n, \mathcal{F}_n)_{n=0,1,...,\infty}$ is also submartingale, where $\mathcal{F}_\infty = \sigma(\cup_{n=1}^\infty \mathcal{F}_n)$.

Proof Statement (a) is obvious. In case (b) one can apply the Fatou lemma and get
$$\mathbf{E} X_\infty = \mathbf{E}\lim_n X_n \geq \limsup_n \mathbf{E} X_n \geq \mathbf{E} X_0 > -\infty.$$

The submartingale property follows from relations:
$$\mathbf{E}(X_\infty | \mathcal{F}_m) = \mathbf{E}(\lim_n X_n | \mathcal{F}_m) \geq \limsup_n E(X_n | \mathcal{F}_m) \geq X_m \text{ (a.s.)}, \quad m = 0, 1, \ldots.$$

□

To study another type of convergence of martingales (submartingales, supermartingales) which is L^1-convergence we give the following example.

6.3 Martingales on infinite time interval

Example 6.1 Let $(Y_n)_{n=0,1,\ldots}$ be a sequence of independent random variables such that
$$Y_n = \begin{cases} 2 & \text{with probability } 1/2, \\ 0 & \text{with probability } 1/2. \end{cases}$$

Define $X_n = \prod_{i=0}^{n} Y_i$ and $\mathcal{F}_n = \sigma(Y_0, \ldots, Y_n)$. It is clear that $(X_n)_{n=0,1,\ldots}$ is a martingale with $\mathbb{E}X_n = 1$, and hence, $X_n \to X_\infty = 0$ (a.s.).

At the same time we have $\mathbb{E}|X_n - X_\infty| = \mathbb{E}X_n = 1$ and therefore, this martingale does not converge in space L^1.

It means that the condition $\sup_n \mathbb{E}|X_n| < \infty$, appeared in Doob's theorem about (a.s.)-convergence, is not enough to achieve L^1-convergence.

Theorem 6.7 *If $(X_n)_{n=0,1,\ldots}$ is a uniformly integrable submartingale, then there exists an integrable random variable X_∞ such that $X_n \xrightarrow[n \to \infty]{} X_\infty$ (a.s.) and in space L^1.*

Moreover, $(X_n)_{n=0,1,\ldots,\infty}$ will be a submartingale with respect to $(\mathcal{F}_n)_{n=0,1,\ldots,\infty}$, where $\mathcal{F}_\infty = \sigma(\cup_{n=1}^{\infty} \mathcal{F}_n)$.

Proof Existence of X_∞ and convergence (a.s.) follows from Theorem 6.6. Further, take $c > 0$ and represent
$$\mathbb{E}X_n = \mathbb{E}X_n I_{\{X_n < -c\}} + \mathbb{E}X_n I_{\{X_n \geq -c\}}. \tag{6.33}$$

For any $\epsilon > 0$ due to uniform integrability (X_n) we can choose $c > 0$ large enough to provide inequality
$$\sup_n |\mathbb{E}X_n I_{\{X_n < -c\}}| < \epsilon, \tag{6.34}$$

for the first term in (6.33).

We also note that
$$X_n I_{\{X_n \geq -c\}} \geq X_n,$$

and by the Fatou lemma we obtain that
$$\liminf_n \mathbb{E}X_n I_{\{X_n \geq -c\}} \geq \mathbb{E}\liminf_n X_n I_{\{X_n \geq -c\}} \geq \mathbb{E}\liminf_n X_n.$$

Hence, due to (6.34) we get from the above inequalities
$$\liminf_n \mathbb{E}X_n \geq \mathbb{E}\liminf_n X_n - \epsilon.$$

We can prove in a similar way
$$\limsup_n \mathbb{E}X_n \leq \mathbb{E}\limsup_n X_n,$$

and further just repeat the same steps of the proof of the Lebesgue dominated convergence theorem and find that $\mathbb{E}|X_n - X_\infty| \to 0$, $n \to \infty$.

To finish the proof we take $A \in \mathcal{F}_n$, and for $m \geq n$ we have
$$\mathbb{E}I_A |X_m - X_\infty| \to 0, \ m \to \infty,$$

$$\int_A X_m dP \to \int_A X_\infty dP, \ m \to \infty,$$

$\{\int_A X_m dP\}_{m \geq n}$ is non-decreasing.

Hence, for any $A \in \mathcal{F}_n$ we have

$$\int_A X_n dP \leq \int_A X_m dP \leq \int_A X_\infty dP = \int_A \mathbf{E}(X_\infty | \mathcal{F}_n) dP,$$

which certifies a submartingale property. □

Corollary 6.3 *Let $(X_n)_{n=0,1,...}$ be a submartingale with $\sup_n \mathbf{E}|X_n|^p < \infty$, $p > 1$. Then there exists an integrable random variable X_∞ such that $X_n \to X_\infty$, $n \to \infty$, (a.s.) and in L^1.*

Theorem 6.8 *(Levy's theorem). Consider an integrable random variable X on stochastic basis $(\Omega, \mathcal{F}, (\mathcal{F})_{n=0,1,...}, P)$. Then*

$$\mathbf{E}(X|\mathcal{F}_n) \xrightarrow[n \to \infty]{} \mathbf{E}(X|\mathcal{F}_\infty) \ (a.s.) \text{ and in } L^1.$$

Proof Denote $X_n = \mathbf{E}(X|\mathcal{F}_n)$, $n = 0, 1, \ldots$ and find for $a > 0$ and $b > 0$ that

$$\int_{\{|X_n| \geq a\}} |X_n| dP \leq \int_{\{|X_n| \geq a\}} \mathbf{E}(|X_n| | \mathcal{F}_n) dP = \int_{\{|X_n| \geq a\}} |X| dP$$

$$= \int_{\{|X_n| \geq a\} \cap \{|X_n| \leq b\}} |X| dP + \int_{\{|X_n| \geq a\} \cap \{|X_n| > b\}} |X| dP$$

$$\leq b P(|X_n| \geq a) + \int_{\{\{|X_n| > b\}} |X| dP$$

$$\leq \frac{b}{a} \mathbf{E}|X| + \int_{\{\{|X_n| > b\}} |X| dP.$$

So, taking $b \to \infty$ and after $a \to \infty$ we get

$$\lim_{a \to \infty} \sup_n \mathbf{E}|X_n| I_{\{|X_n| \geq a\}} = 0,$$

which means that $(X_n)_{n=0,1,...}$ is uniformly integrable. Applying Theorem 6.7 we get the statement of this theorem. □

Corollary 6.4 *A uniformly integrable stochastic sequence $(X_n)_{n=0,1,...}$ on stochastic basis $(\Omega, \mathcal{F}, (\mathcal{F}_n)_{n=0,1,...}, P)$ is a martingale $\Leftrightarrow$ there exists an integrable random variable X such that $X_n = \mathbf{E}(X|\mathcal{F}_n)$, $n = 0, 1, \ldots$.*

Proof The inverse implication $\Leftarrow$ is just the Levy theorem (Theorem 6.8). The direct implication $\Rightarrow$ is just a consequence of Theorem 6.7 if we take $X = X_\infty$. □

Problem 6.2 In Levy's theorem prove that

$$X_\infty = \mathbf{E}(X|\mathcal{F}_\infty).$$

Chapter 7
Discrete time stochastic analysis: further results and applications

Abstract In this chapter a characterization of sets of convergence of martingale is given in predictable terms. As a consequence, the strong LNL for square-integrable martingales is proved. This result is applied for derivation of strong consistency of the least-squared estimates in the framework of regression model with martingale errors. Moreover, the CLT for martingales is stated, and further this theorem together with the martingale LNL is applied to derive the asymptotic normality and strong consistency of martingale stochastic approximation procedures. A discrete version of the Girsanov theorem is given here with its further application for derivation of a discrete time Bachelier option pricing formula. In the last section, the notion of a martingale is extended in several directions: from asymptotic martingales and local martingales to martingale transforms and generalized martingales (see [4], [8], [12], [13], [15], [26], [30], [34], and [40]).

7.1 Limiting behavior of martingales with statistical applications

Let us investigate limiting behavior of martingales. We start with some facts about sets of convergence of martingales and submartingales.

Definition 7.1 Let $(X_n)_{n=0,1,\ldots}$ be a stochastic sequence. Denote $\{\omega : X_n \to\}$ the set of those $\omega \in \Omega$ such that $X_n(\omega)$ converges to a finite limit as $n \to \infty$.

We also say that $A \subseteq B \in \mathcal{F}$ (a.s.), if $P(A \cap B^c) = 0$, and $A = B$ (a.s.) if $A \subseteq B$ (a.s.) and $B \subseteq A$ (a.s.).

Lemma 7.1 Let $(X_n)_{n=0,1,\ldots}$ be a square integrable martingale. Then (a.s.)

$$\{\omega : \langle X, X \rangle_\infty < \infty\} \subseteq \{\omega : X_n \to\} \tag{7.1}$$

Proof For a positive $a \in \mathbb{R}$ we define a stopping time

$$\tau_a = \inf\{n : \langle X, X \rangle_{n+1} \geq a\},$$

assuming that $\inf\{\emptyset\} = \infty$. One can observe that $\langle X, X\rangle_{\tau_a} \leq a$ and the stopped sequence $(X_n^{\tau_a})_{n=0,1,\ldots} = (X_{\tau_a \wedge n})_{n=0,1,\ldots}$ is a square integrable martingale such that

$$\mathbf{E}(X_{\tau_a \wedge n})^2 = \mathbf{E}(X_n^{\tau_a})^2 = \mathbf{E}\langle X^{\tau_n}, X^{\tau_n}\rangle_n = \mathbf{E}\langle X, X\rangle_{\tau_a \wedge n} \leq a.$$

According to Theorem 6.6, $\mathbf{P}\{\omega : X_n^{\tau_n} \to\} = 1$. We also note that $X_n^{\tau_n} = X_{\tau_a \wedge n} = X_n$ on the set $\{\tau_a = \infty\} = \{\omega : \langle X, X\rangle_\infty \leq a\}$.

Further, (a.s.) $\{\omega : \tau_a = \infty\} \subseteq \{\omega : X_n \to\}$ and $\cup_{a>0}\{\tau_a = \infty\} \subseteq \{\omega : X_n \to\}$. Hence, (a.s.)

$$\{\omega : \langle X, X\rangle_\infty < \infty\} = \cup_{a>0}\{\omega : \langle X, X\rangle_\infty \leq a\} = \cup_{a>0}\{\omega : \tau_a = \infty\} \subseteq \{\omega : X_n \to\},$$

and we get (7.1). □

Lemma 7.2 *(Stochastic Kronecker's lemma). Let* $(A_n)_{n=0,1,\ldots}$, $A_0 > 0$, *be a predictable non-decreasing sequence,* $(M_n)_{n=0,1,\ldots}$ *be a square integrable martingale and*

$$N_n = \sum_{i=0}^{n} A_i^{-1} \Delta M_i.$$

Then (a.s.)

$$\Omega' = \{\omega : A_\infty = \infty\} \cap \{\omega : N_n \to\} \subseteq \{\omega : A_n^{-1} M_n \to\}. \quad (7.2)$$

Proof First of all, we note the following formula of "summation by parts":

$$A_n X_n = A_0 X_0 + \sum_{i=1}^{n} A_i \Delta X_i + \sum_{i=1}^{n} X_{i-1} \Delta A_i, \quad (7.3)$$

for two stochastic sequences $(A_n)_{n=0,1,\ldots}$ and $(X_n)_{n=0,1,\ldots}$.

Applying the formula (7.3) to $(A_n)_{n=0,1,\ldots}$ and $(N_n)_{n=0,1,\ldots}$ we represent

$$\sum_{i=1}^{n} A_i \Delta N_i = A_n N_n - \sum_{i=1}^{n} N_{i-1} \Delta A_i,$$

and using $\sum_{i=1}^{n} A_i \Delta N_i = M_n$ we obtain

$$A_n^{-1} M_n = A_n^{-1} \sum_{i=1}^{n} A_i \Delta N_i = A_n^{-1} \left[A_n N_n - \sum_{i=1}^{n} N_{i-1} \Delta A_i \right] = A_n^{-1} \sum_{i=1}^{n} (N_n - N_{i-1}) \Delta A_i. \quad (7.4)$$

Further, for $\epsilon > 0$, define

$$n_\epsilon = \begin{cases} \sup\{i : |N_\infty - N_{i-1}| \leq \epsilon\} & \text{on the set } \{N_n \to N_\infty\}, \\ 0, & \text{on } \Omega \setminus \{N_n \to N_\infty\}. \end{cases}$$

7.1 Limiting behavior of martingales with statistical applications

By (7.4) on the set $\{\omega : N_n \to N_\infty\}$ we have

$$
\begin{aligned}
|A_n^{-1} M_n| &= |A_n^{-1} \sum_{i=1}^{n} (N_n - N_{i-1}) \Delta A_i| \\
&\leq A_n^{-1} | \sum_{i=1}^{n} (N_n - N_{i-1}) \Delta A_i| \\
&\leq A_n^{-1} \left[\sum_{i=1}^{n \wedge n_\epsilon} |N_n - N_{i-1}| \Delta A_i + \sum_{i=n \wedge n_\epsilon}^{n} [|N_\infty - N_n| + |N_\infty - N_{i-1}|] \Delta A_i \right] \\
&\leq A_n^{-1} \left[const. A_{n_\epsilon} \sup_i |N_i| + const. A_n |N_\infty - N_n| + const. \epsilon A_n \right]. \quad (7.5)
\end{aligned}
$$

Hence, for almost all $\omega \in \Omega'$ one can find $n'_\epsilon(\omega)$ such that in (7.5) for $n \geq n'_\epsilon(\omega)$: $|A_n^{-1} M_n(\omega)| \leq const.(\omega) \cdot \epsilon$, and therefore (7.2). □

Theorem 7.1 (*Strong LNL for martingales*). *Let $(A_n)_{n=0,1,...}$ be a non-negative predictable non-decreasing sequence and $(M_n)_{n=0,1,...}$, $M_0 = 0$, be a square integrable martingale such that (a.s.) $A_\infty = \infty$ and (a.s.)*

$$\sum_{n=1}^{\infty} \frac{\Delta \langle M, M \rangle_n}{A_n^2} < \infty. \quad (7.6)$$

Then $A_n^{-1} M_n \to 0$ (a.s.), $n \to \infty$.

Proof Due to assumptions (7.6) and Lemma 7.1 we observe in Lemma 7.2 $P(\Omega') = 1$, and, hence, we get the statement of the theorem. □

Corollary 7.1 (*Strong LNL of Kolmogorov*). *Let $(Y_n)_{n=1,2,...}$ be a sequence of independent random variables with $\mathbf{E} Y_n = 0$ and $\sigma_n^2 = \mathbf{E} Y_n^2$ such that*

$$\sum_{n=1}^{\infty} \frac{\sigma_n^2}{n^2} < \infty. \quad (7.7)$$

Then for a sequence $X_n = \sum_{i=1}^{n} Y_i$ the following strong large numbers law is true: (a.s.) $\frac{X_n}{n} \to 0$, $n \to \infty$.

Proof We put $A_n = n$, $M_n = X_n$ and find that the *Kolmogorov variance condition* (7.7) is transformed to the condition (7.6) of Theorem 7.1 and we get the claim. □

Let us give an interesting and valuable *application* of strong LNL for martingales to *Regression Analysis*.

Example 7.1 Suppose the observations are performed at times $n = 0, 1, \ldots$ and obey the formula

$$x_n = f_n \theta + e_n, \quad (7.8)$$

where $\theta \in \mathbb{R}$ is an unknown parameter, $(e_n)_{n=0,1,\ldots}$ is a martingale-difference and $(f_n)_{n=0,1,\ldots}$ is a predictable regressor.

One can consider a *structural least squares estimate (LS-estimate)* in the framework of regression model (7.8):

$$\theta_n = \left(\sum_{i=0}^n f_i^2\right)^{-1} \sum_{i=0}^n f_i x_i. \tag{7.9}$$

Denote $D_n = \mathbf{E}(e_n^2|\mathcal{F}_{n-1})$ and $F_n = \sum_{i=0}^n f_i^2$ and assume that $F_n \to \infty$ (a.s.), $n \to \infty$, and $\sum_{i=0}^\infty F_i^{-2} D_i < \infty$ (a.s.). Then according to Theorem 7.1 LS-estimate (7.9) is *strongly consistent* in the sense that $\theta_n \to \theta$ (a.s.), $n \to \infty$.

Remark 7.1 If in the model (7.8) we take $f_n = x_{n-1}$, then we arrive to the first order autoregression model for which the condition $\sum_1^\infty x_{i-1}^2 = \infty$ is well-known in Regression Analysis as a standard guarantee for consistency of LS-estimates.

Below we give some additional facts about asymptotic behavior of martingales and submartingales.

Definition 7.2 A stochastic sequence $(X_n)_{n=0,1,\ldots} \in C^+$, if for each $\tau_a = \inf\{n \geq 0 : X_n > a\}$, $a > 0$:

$$\mathbf{E}(\Delta X_{\tau_a})^+ I_{\{\omega:\tau_a < \infty\}} < \infty.$$

Problem 7.1 Prove that the condition C^+ is true for any stochastic sequence with $\mathbf{E}\sup_n |\Delta X_n| < \infty$.

Theorem 7.2 *a) Assume a submartingale with its Doob decomposition $X_n = X_0 + A_n + M_n$, $n = 0, 1, \ldots$ belongs to a class C^+, then (a.s.)*

$$\{X_n \to\} = \{\sup_n X_n < \infty\} \subseteq \{A_\infty < \infty\}.$$

b) Assume a submartingale $X_n = X_0 + A_n + M_n$, $n = 0, 1, \ldots$ is non-negative, then (a.s.)

$$\{A_\infty < \infty\} \subseteq \{X_n \to\} \subseteq \{\sup_n X_n < \infty\}.$$

c) If a non-negative submartingale belongs to C^+, then (a.s.)

$$\{A_\infty < \infty\} = \{X_n \to\} = \{\sup_n X_n < \infty\}.$$

d) Let $(Y_n)_{n=0,1,\ldots}$ be a non-negative stochastic sequence with the structure

$$Y_n = Y_0 + A'_n + M_n - A_n^2, \ n = 0, 1, \ldots,$$

where $(A'_n)_{n=0,1,\ldots}$ and $(A_n^2)_{n=0,1,\ldots}$ are non-decreasing predictable sequences and $(M_n)_{n=0,1,\ldots}$ is a martingale. Then (a.s.)

$$\{A'_\infty < \infty\} \subseteq \{Y_n \to\} = \{\sup_n A_\infty^2 < \infty\}.$$

7.1 Limiting behavior of martingales with statistical applications

Proof a) It is clear that (a.s.) $\{X_n \to\} \subseteq \{\sup_n X_n < \infty\}$. To prove the inverse inclusion, we have

$$\sup_n EX^+_{\tau_a \wedge n} \leq a + EX^+_{\tau_a} I_{\tau_a < \infty} \leq 2a + E(\Delta X_{\tau_a})^+ I_{\{\tau_a < \infty\}} < \infty.$$

Hence, the stopped submartingale $(X_n^{\tau_a})_{n=0,1,\ldots} = (X_{\tau_a \wedge n})_{n=0,1,\ldots}$ converges (a.s.) as $n \to \infty$, and $\{\tau_a = \infty\} \subseteq \{X_n \to\}$ (a.s.).
Due to $\cup_{a>0}\{\tau_a = \infty\} = \{\sup_n X_n < \infty\}$ we get

$$\{\sup_n X_n < \infty\} = \{X_n \to\} \ (a.s.).$$

b) Define $\sigma_a = \inf(n : A_{n+1} > a), a > 0$, and get

$$EX_{n \wedge \sigma_a} = EA_{n \wedge \sigma_a} \leq a.$$

Hence $\sup_n EX_{n \wedge \sigma_a} \leq a < \infty$, and due to positivity of $(X_n)_{n=0,1,\ldots}$ we can apply the Doob convergence theorem and find that

$$\{A_\infty \leq a\} \subseteq \{\sigma_a = \infty\} = \{X_n \to\} \ (a.s.)$$

and therefore,

$$\{A_\infty < \infty\} = \cup_{a>0}\{A_\infty \leq a\} \subseteq \{X_n \to\} \ (a.s.).$$

c) is a combination of (a) and (b).

d) Let us rewrite $Y_n = X_n - A_n^2$, $n = 0, 1, \ldots$ where $X_n = X_0 + A'_n + M_n$. $X_0 = Y_0$, is a non-negative submartingale due to $0 \leq Y_n = X_n - A_n^2$, $0 \leq A_n^2 \leq X_n$ (a.s.), and (a.s.) $\{A'_\infty < \infty\} \subseteq \{X_n \to\} \subseteq \{A_\infty^2 < \infty\}$ by (b). Hence, (a.s.) $\{A'_\infty < \infty\} \subseteq \{Y_n \to\} \cap \{A_\infty^2 < \infty\}$. □

Corollary 7.2 *Assume a martingale* $(X_n)_{n=0,1,\ldots}$ *satisfies to the following condition* $E \sup_n |\Delta X_n| < \infty$. *Then (a.s.)*

$$\Omega = \{X_n \to\} \cup \{\liminf_n X_n = -\infty, \limsup_n X_n = +\infty\}. \tag{7.10}$$

Proof We apply Theorem 7.2 to $\pm(X_n)_{n=0,1,\ldots}$ and get (a.s.)

$$\{\limsup_n X_n < \infty\} = \{\sup_n X_n < \infty\} = \{X_n \to\},$$

$$\{\liminf_n X_n > -\infty\} = \{\inf_n X_n > -\infty\} = \{X_n \to\},$$

which leads to (7.10). □

We already studied the Large Numbers Law for martingales and a very nice generalization of the corresponding results for sums of independent random variables observed.
A natural question arises.

Is it possible to get a similar version of the *Central Limit Theorem (CLT) for martingales?*

The next theorem gives a positive answer to this question.

Theorem 7.3 *Let* $(X_n)_{n=0,1,...}$ *be a martingale-difference such that (a.s.)* $\mathbf{E}(X_n^2|\mathcal{F}_{n-1}) = 1$ *and* $\mathbf{E}(|X_n|^3|\mathcal{F}_{n-1}) \leq C < \infty$. *Then for a martingale* $M_n = \sum_{i=1}^n X_i$ *the CLT is true:*

$$Y_n = \frac{M_n}{\sqrt{n}} \xrightarrow[n\to\infty]{d} Z \sim N(0,1)$$

Proof Here, as in Chapter 4 for a random variable X we define its characteristic function $\phi_X(t) = \mathbf{E}e^{itX}$, $t \in \mathbb{R}$, and denote

$$\phi_{n,j}(t) = \mathbf{E}\left(e^{it\frac{X_j}{\sqrt{n}}}|\mathcal{F}_{j-1}\right).$$

Let us write the Taylor decomposition

$$e^{it\frac{X_j}{\sqrt{n}}} = 1 + it\frac{X_j}{\sqrt{n}} - \frac{t^2}{2n}X_j^2 - \frac{it^3}{6n^{3/2}}\bar{X}_j^3, \qquad (7.11)$$

with a random reminder bounded as follows $0 \leq \bar{X}_j \leq |X_j|$.

Taking the conditional expectation in (7.11) and exploiting a martingale-difference assumption we obtain the corresponding representation for $\phi_{n,j}(t)$:

$$\phi_{n,j}(t) = 1 + it\frac{\mathbf{E}(X_j|\mathcal{F}_{j-1})}{\sqrt{n}} - \frac{t^2}{2n}\mathbf{E}(X_j^2|\mathcal{F}_{j-1}) - \frac{it^3}{6n^{3/2}}\mathbf{E}(\bar{X}_j^3|\mathcal{F}_{j-1}). \qquad (7.12)$$

It follows from (7.12) that

$$\phi_{n,j}(t) - 1 - \frac{t^2}{2n} = \frac{t^3}{6n^{3/2}}\mathbf{E}(\bar{X}_j^3|\mathcal{F}_{j-1}).$$

Hence, for $m \leq n$ we obtain from (7.10) that

$$\mathbf{E}e^{it\frac{M_m}{\sqrt{n}}} = \mathbf{E}e^{it\frac{M_{m-1}}{\sqrt{n}}}e^{it\frac{X_m}{\sqrt{n}}} = \mathbf{E}\left(e^{it\frac{M_{m-1}}{\sqrt{n}}}\mathbf{E}\left(e^{it\frac{X_m}{\sqrt{n}}}|\mathcal{F}_{m-1}\right)\right)$$

$$= \mathbf{E}e^{it\frac{M_{m-1}}{\sqrt{n}}}\phi_{n,m}(t) \qquad (7.13)$$

$$= \mathbf{E}e^{it\frac{M_{m-1}}{\sqrt{n}}}\left(1 - \frac{t^2}{2n} - \frac{it^3}{6n^{3/2}}\bar{X}_j^3\right).$$

Let us rewrite (7.11) as follows

$$\mathbf{E}\left(e^{it\frac{M_m}{\sqrt{n}}} - (1 - \frac{t^2}{2n})e^{it\frac{M_{m-1}}{\sqrt{n}}}\right) = \mathbf{E}e^{it\frac{M_{m-1}}{\sqrt{n}}}\frac{it^3}{6n^{3/2}}\bar{X}_j^3. \qquad (7.14)$$

Using the boundedness of the third moment of $(X_n)_{n=0,1,...}$ we derive from (7.13) that

7.1 Limiting behavior of martingales with statistical applications

$$\left| \mathbf{E}\left(e^{it\frac{M_m}{\sqrt{n}}} - (1 - \frac{t^2}{2n})e^{it\frac{M_{m-1}}{\sqrt{n}}} \right) \right|$$

$$\leq \mathbf{E}\left(\left| e^{it\frac{M_{m-1}}{\sqrt{n}}} \right| \frac{|t|^3}{6n^{3/2}} \mathbf{E}\left(|X_m|^3 | \mathcal{F}_{m-1} \right) \right) \leq c\frac{|t|^3}{6n^{3/2}}. \quad (7.15)$$

Let us fix $t \in \mathbb{R}$ and choose large enough $n \geq \frac{t^2}{2}$ to provide the following inequality $0 \leq 1 - t^2/2n \leq 1$.

For such t and n we obtain from (7.13)-(7.15) that

$$\left| (1 - \frac{t^2}{2n})^{n-m} \mathbf{E}e^{it\frac{M_{m-1}}{\sqrt{n}}} - (1 - \frac{t^2}{2n})^{n-m+1} \mathbf{E}e^{it\frac{M_{m-1}}{\sqrt{n}}} \right| \leq c\frac{|t|^3}{6n^{3/2}}. \quad (7.16)$$

Let us note that

$$\mathbf{E}e^{it\frac{M_m}{\sqrt{n}}} - (1 - \frac{t^2}{2n})^n = \sum_{m=1}^{n} \left[(1 - \frac{t^2}{2n})^{n-m} \mathbf{E}e^{it\frac{M_{m-1}}{\sqrt{n}}} - (1 - \frac{t^2}{2n})^{n-m+1} \mathbf{E}e^{it\frac{M_{m-1}}{\sqrt{n}}} \right]. \quad (7.17)$$

Relations (7.16)-(7.17) for $n \geq t^2/2$ lead to the following inequality

$$\left| \mathbf{E}e^{it\frac{M_m}{\sqrt{n}}} - (1 - \frac{t^2}{2n})^n \right| \leq nC\frac{|t|^3}{6n^{3/2}} = C\frac{|t|^3}{6\sqrt{n}}. \quad (7.18)$$

Due to the right hand side of (7.18) tends to zero, now one can conclude that

$$\lim_{n \to \infty} \mathbf{E}e^{it\frac{M_m}{\sqrt{n}}} = e^{-t^2/2}, \quad (7.19)$$

where we used the well-known fact that

$$\lim_{n \to \infty} \left(1 - \frac{t^2}{2n} \right)^n = e^{-t^2/2}.$$

Relation (7.19) shows that characteristic functions of Y_n converge to the characteristic function of $N(0, 1)$. Hence, $Y_n \xrightarrow[n \to \infty]{d} Z \sim N(0, 1)$. □

Another area of applications of a martingale LNL and a martingale CLT is stochastic approximation algorithms. The classical theory of stochastic approximation is concerned with the problem to construct a stochastic sequence $(\theta_n)_{n=0,1,\ldots}$ that converges in some probabilistic sense to a unique root θ of the regression equation

$$R(\theta) = 0, \ \theta \in \mathbb{R}, \quad (7.20)$$

where R is a regression function.

An approximate solution of (7.20) is given by the Robbins-Monro procedure

$$\theta_n = \theta_{n-1} - \gamma_n y_n, \quad (7.21)$$

with the sequence (y_n) satisfying to equation

$$y_n = R(\theta_{n-1}) + e_n, \tag{7.22}$$

where $(e_n)_{n=0,1,...}$ are errors of observations usually modeled by a sequence of independent random variables with mean zero and bounded variance, $(\gamma_n)_{n=0,1,...}$ is a sequence of positive real numbers converging to 0.

The convergence of the procedure (7.21)-(7.22) for a continuous linearly bounded regression function $R(x)$ with $R(x)(x - \theta) > 0$ for all $x \in \mathbb{R}$ is guaranteed by the following classical conditions:

$$\sum_{1}^{\infty} \gamma_k = \infty, \tag{7.23}$$

$$\sum_{1}^{\infty} \gamma_k^2 < \infty. \tag{7.24}$$

Let us demonstrate how the martingale technique and convergence work here and how classical stochastic approximation results are extended to models (7.21)-(7.22) with martingale errors $(e_n)_{n=0,1,...}$ and predictable sequences $(\gamma_n)_{n=0,1,...}$.

Example 7.2 We consider for simplicity the case of linear function $R(x) = \beta(x - \theta)$, $\beta > 0$, assuming in (7.22) that $(e_n)_{n=0,1,...}$ is a martingale-difference with $\mathbf{E}e_n = 0$ and $\mathbf{E}(e_n^2|\mathcal{F}_{n-1}) \leq \xi < \infty$ (a.s.). Regarding (γ_n) we assume that (γ_n) is predictable, $0 < \gamma_n \leq \beta^{-1}$ (a.s.) and conditions (7.23)-(7.24) are fulfilled almost surely. One can rewrite the algorithm (7.21)-(7.22) as follows ($\theta = 0$ for simplicity)

$$\Delta \theta_n = -\gamma_n \beta \theta_{n-1} - \gamma_n e_n. \tag{7.25}$$

Moreover, (7.25) can be rewritten in the form of an inhomogeneous linear stochastic differential equation (6.9) with $X_n = \theta_n$, $\Delta N_n = -\gamma_n e_n$, $\Delta U_n = -\beta \gamma_n$, and $\mathcal{E}_n(-\beta\gamma)$ is a stochastic exponential of sequence (U_n).

In these denotations the solution of (7.25) is expressed as follows

$$\theta_n = X_n = \mathcal{E}_n(U)X_0 + \mathcal{E}_n(U)\sum_{1}^{n} \mathcal{E}_i^{-1}(U)\Delta N_i$$

$$= \mathcal{E}_n(-\beta\gamma)\theta_0 + \mathcal{E}_n(-\beta\gamma)\sum_{1}^{n} \mathcal{E}_i^{-1}(-\beta\gamma)\gamma_i e_i. \tag{7.26}$$

According to the assumption $\gamma_n < \beta^{-1}$ (a.s.) the stochastic exponential $\mathcal{E}_n(-\beta\gamma) = \prod_{1}^{n}(1 - \beta\gamma_i)$ is positive (a.s.) if (7.23) is fulfilled (a.s.).

Further, the first term of the right hand side of (7.26) tends to zero (a.s.), $n \to \infty$, because $\mathcal{E}_n(-\beta\gamma) \to 0$ (a.s.), $n \to \infty$.

The structure of the second term of the right hand side of (7.26) is exactly (see Theorem 7.1) as in the strong LNL for martingales with $A_n^{-1} = \mathcal{E}_n(-\beta\gamma)$ and $M_n = \sum_{1}^{n} \mathcal{E}_i^{-1}(-\beta\gamma)\gamma_i e_i$.

7.1 Limiting behavior of martingales with statistical applications

So, we have to check the condition (7.6) in this case:

$$\sum_1^\infty A_n^{-2}\Delta\langle M,M\rangle_n = \sum_1^n \mathcal{E}_n^2(-\beta\gamma)\mathcal{E}_n^{-2}(-\beta\gamma)\gamma_n^2 \mathrm{E}(e_n^2|\mathcal{F}_{n-1}) \le \xi \sum_1^\infty \gamma_n^2 < \infty \quad (a.s.)$$

Applying Theorem 7.1 we get the convergence $\theta_n \to \theta$ (a.s.), $n \to \infty$.

Let us show how the martingale CLT (see Theorem 7.3) works to study asymptotic normality properties of algorithms (7.21)-(7.22). Under some reasonable simplifications, we will show it to avoid technical difficulties.

Example 7.3 We consider the linear model (7.21)-(7.22) assuming $\gamma_n = \frac{\alpha}{n}$, $\alpha > 0$, $e_n \sim N(0, \sigma^2)$ and independent, and n must be greater than $\alpha\beta$ to provide a positivity of $\mathcal{E}_n(-\beta\gamma)$.

Problem 7.2 Prove that under the assumptions above

$$\mathcal{E}_n(-\beta\gamma) \equiv \mathcal{E}_n(-\beta\alpha) \sim n^{-\beta\alpha}, \quad n \to \infty. \tag{7.27}$$

To derive an asymptotic normality of the procedure (θ_n) we multiply (7.26) by $n^{1/2}$ and obtain

$$\sqrt{n}\theta_n = \sqrt{n}\mathcal{E}_n(-\beta\alpha)\theta_0 - \sqrt{n}\mathcal{E}_n(-\beta\alpha)\sum_1^n \frac{\alpha}{k}\mathcal{E}_k^{-1}(-\beta\alpha)e_k. \tag{7.28}$$

It follows from (7.27) that $\sqrt{n}\mathcal{E}_n(-\beta\alpha) \sim n^{1/2-\beta\alpha}$, and, hence, under additional assumption $2\beta\alpha > 1$ we find the first term of the right side of (7.28) tends to zero (a.s.), $n \to \infty$. The second term of (7.28) has a normal distribution. Therefore, we need to calculate its limiting variance.

As $n \to \infty$ we have

$$n\mathcal{E}_n^2(-\beta\alpha)\sum_1^n \mathcal{E}_k^{-2}(-\beta\alpha)\frac{\alpha^2\sigma^2}{k^2} \to \frac{\alpha^2\sigma^2}{(2\beta\alpha-1)}. \tag{7.29}$$

Thus we obtain from (7.29) that

$$\sqrt{n}\theta_n \xrightarrow[n\to\infty]{d} N\left(0, \frac{\alpha^2\sigma^2}{2\beta\alpha-1}\right).$$

Remark 7.2 One can develop this approach to a non-linear regression functions $R(x) = \beta(x - \theta) + U(x)$, where $U(x) = O((x-\theta)^2)$, and prove both strong consistency and asymptotic normality of stochastic approximation procedures (7.21)-(7.22). We note also that Lemma 7.2 works for such extension.

7.2 Martingales and absolute continuity of measures. Discrete time Girsanov theorem and its financial application

We start this section with a martingale characterization of absolute continuity of a probability measure $\tilde{P}$ with respect to measure P on given stochastic basis $(\Omega, \mathcal{F}, (\mathcal{F}_n)_{n=0,1,\ldots}, P)$. It was already shown in Section 6.1 that under assumption of "local absolute continuity" $\tilde{P} \ll^{loc} P$ the corresponding local density $Z_n = \frac{d\tilde{P}_n}{dP_n}$ is a martingale w.r. to P. The next theorem states conditions under which a "local absolute continuity" is transformed to "absolute continuity" of $\tilde{P}$ w.r. to P.

Theorem 7.4 *Assume that* $\tilde{P} \ll_{loc} P$ *and* $(Z_n)_{n=0,1,\ldots}$, $Z_0 = 1$, *is a local density* $\frac{d\tilde{P}_n}{dP_n}$. *Then the following statements are equivalent:*
1) $\tilde{P} \ll P$,
2) $(Z_n)_{n=0,1,\ldots}$ *is uniformly integrable,*
3) $\tilde{P}(\omega : \sup_n Z_n < \infty) = 1$.

Proof (1) $\Rightarrow$ (3): According to Theorem 6.6 there exists $Z_\infty = \lim_{n\to\infty} Z_n$ P-a.s. Due to $\tilde{P} \ll P$ such a limit does exist $\tilde{P}$-a.s.. Hence, $\tilde{P}(\omega : \sup_n Z_n < \infty) = 1$.

(3) $\Rightarrow$ (2): For a constant $c > 0$ we have

$$\mathbf{E} Z_n I_{\{\omega : Z_n > c\}} = \tilde{P}(\omega : Z_n > c) \le \tilde{P}(\omega : \sup_i Z_i > c) \to 0, \ c \to \infty.$$

(2) $\Rightarrow$ (1): Due to the uniform integrability of $(Z_n)_{n=0,1,\ldots}$ we can apply Theorem 6.7 and find the existence $Z_\infty = \lim_{n\to\infty} Z_n$ P-a.s., and $\mathbf{E}|Z_n - Z_\infty| \to 0$, $n \to \infty$.

Further, for any $A \in \mathcal{F}_m$ and for $n \ge m$, we can write a martingale property of $(Z_n)_{n=0,1,\ldots}$ as follows

$$\tilde{P}(A) = \mathbf{E} I_A Z_m = \mathbf{E} I_A Z_n. \tag{7.30}$$

It follows from L^1-convergence that one can take a limit as $n \to \infty$ in the equality (7.30) and obtain

$$\tilde{P}(A) = \mathbf{E} I_A Z_\infty,$$

i.e. $Z_\infty = \frac{d\tilde{P}}{dP}$. □

Now we show a method of construction of probability measure $\tilde{P}$ and its (local) density w.r. to P. Aiming it we formulate one of the simplest versions of the *Girsanov theorem*.

Suppose $(\epsilon_n)_{n=0,1,\ldots}$ is a sequence of independent random variables $\epsilon_n \sim N(0, 1)$. Define $\mathcal{F}_0 = \{\emptyset, \Omega\}$, $\mathcal{F}_n = \sigma(\epsilon_1, \ldots, \epsilon_n)$, $n = 1, 2, \ldots$, a bounded predictable sequence $(\alpha_n)_{n=0,1,\ldots}$ and $Z_0 = 1$,

$$Z_n = \exp\left\{-\sum_1^n \alpha_k \epsilon_k - \frac{1}{2}\sum_1^n \alpha_k^2\right\}, \ n = 1, 2, \ldots \tag{7.31}$$

Problem 7.3 Prove that $(Z_n)_{n=0,1,\ldots}$ is a martingle with $\mathbf{E} Z_n = 1$.

7.2 Martingales and absolute continuity of measures. Discrete time ...

Theorem 7.5 *Define a new probability measure* $\tilde{P}_N(A) = \mathbf{E}I_A Z_N$, $A \in \mathcal{F}_N$. *Then* $\tilde{\epsilon}_n = \alpha_n + \epsilon_n$, $n = 1, \ldots, N$, *is a sequence of independent standard normal random variables w.r. to* $\tilde{P}_N$.

Proof For real numbers $(\lambda_n)_{n=1,\ldots,N}$ we have using the rule of changing of measures, properties of conditional expectations and the form of characteristic function for standard normal random variables:

$$\tilde{\mathbf{E}}_N e^{i \sum_1^N \lambda_n \tilde{\epsilon}_n} = \mathbf{E} Z_N e^{i \sum_1^N \lambda_n \tilde{\epsilon}_n}$$
$$= \mathbf{E}\left[e^{i \sum_1^{N-1} \lambda_n \tilde{\epsilon}_n} Z_{N-1} \mathbf{E}\left(e^{i \lambda_N (\alpha_N + \epsilon_N) + \alpha_N \epsilon_N - \alpha_N^2/2} | \mathcal{F}_{N-1} \right) \right]$$
$$= \mathbf{E}\left[e^{i \sum_1^{N-1} \lambda_n \tilde{\epsilon}_n} Z_{N-1} \right] e^{-\lambda_N^2/2} = \ldots = e^{-1/2 \sum_1^N \lambda_n^2}.$$

It means that characteristic function of $\sum_1^N \lambda_n \tilde{\epsilon}_n$ is a product of characteristic functions of standard normal random variables and we get the statement of the theorem. □

Remark 7.3 One can extend Theorem 7.5 to infinite the time interval. According to Theorem 7.4 we need provide a condition to guarantee that $(Z_n)_{n=0,1,\ldots}$ is uniformly integrable. Usually, it is achieved with the help of the Novikov type condition $\mathbf{E} \exp(1/2 \sum_1^\infty \alpha_n^2) < \infty$.

Let us give an example of application of this theorem to Mathematical Finance.

Example 7.4 We assume that financial market consists of two assets $(B_n)_{n=0,1,\ldots,N}$ and $(S_n)_{n=0,1,\ldots,N}$ (bank account and stock prices relatively). We put for simplicity that $B_n \equiv 1$ (interest rate is zero) and

$$\Delta S_n = S_n - S_{n-1} = \alpha_n + \epsilon_n, S_0 > 0, n = 1, \ldots, N. \tag{7.32}$$

We consider a standard financial contract "call option" with a strike K. The holder of the contract has the right to buy a stock share by the price K at the maturity date N. In terms of having the right - he /she must pay a premium $\mathbb{C}_N$ at time 0. The basic problem here is to determine $\mathbb{C}_N$.

The market (7.32) can be considered as the *discrete time Bachelier model* with unit volatility. It is well-established that the premium must be calculated to avoid *arbitrage*. According to this financial no-arbitrage principle $\mathbb{C}_N$ is determined as expected value of pay-off $(S_N - K)^+$ w.r. to a risk-neutral measure $\tilde{P}_N$ given in the Girsanov theorem.

To provide concrete calculations we formulate an auxiliary fact for a Normal distribution as a problem.

Problem 7.4 Prove that

$$\mathbf{E}(a + b\epsilon)^+ = a\Phi(a/b) - b\phi(a/b), \tag{7.33}$$

where $\epsilon \sim N(0, 1)$, $a \in \mathbb{R}$, $b > 0$, $\phi = \phi(x)$ is a standard normal density and $\Phi(x) = \int_{-\infty}^x \phi(y) dy$.

To calculate $\mathbb{C}_N$ we apply (7.33) and obtain the following discrete time *formula of Bachelier*

$$\mathbb{C}_N = \tilde{\mathbf{E}}_N(S_N - K)^+ = \tilde{\mathbf{E}}_N \left(S_0 + \sum_1^N \alpha_i + \sum_1^n \epsilon_i - K \right)^+$$

$$= \mathbf{E} \left(S_0 - K + \sum_1^N \epsilon_i \right)^+ = \mathbf{E}(S_0 - K + \sqrt{N}\epsilon)^+$$

$$= (S_0 - K)\Phi\left(\frac{S_0 - K}{\sqrt{N}}\right) + \sqrt{N}\phi\left(\frac{S_0 - K}{\sqrt{N}}\right).$$

7.3 Asymptotic martingales and other extensions of martingales

We start with an idea to avoid conditional expected values to extend the notion of martingale. Let us assume that $(\Omega, \mathcal{F}, (\mathcal{F}_n)_{n=0,1,\ldots}, P)$ be a stochastic basis on which all stochastic sequences are considered. Denote T_b the set of all bounded stopping times on this stochastic basis. We defined before that two stopping times $\tau \leq \sigma$ if $\tau(\omega) \leq \sigma(\omega)$ (a.s.). With this definition the set T_b is a *directed set* filtering to the right.

Definition 7.3 A stochastic sequence of integrable random variables $(X_n)_{n=0,1,\ldots}$ is called an **asymptotic martingale (amart)**, if the family (**net**) $(\mathbf{E}X_\tau)_{\tau \in T_b}$ converges.

Remark 7.4 It follows from the definition of amart $(X_n)_{n=0,1,\ldots}$ that the set $(\mathbf{E}X_\tau)_{\tau \in T_b}$ is bounded. It is also clear that a linear combinations of amarts is an amart.

According to this definitions and Theorem 6.6 and Theorem 6.7, we can conclude that martingales belong to the class of asymptotic martingales. In this case a natural question arises:
"What elements of previously developed martingale theory can be extended to a wider class of asymptotic martingales?"
If another filtration $(\mathcal{G}_n)_{n=0,1,\ldots}$ is included to filtration $(\mathcal{F}_n)_{n=0,1,\ldots}$ i.e. $\mathcal{G}_n \subseteq \mathcal{F}_n$, $n = 0, 1, \ldots$, then each stopping time w.r. to $(\mathcal{G}_n)$ will be a stopping time w.r. to $(\mathcal{F}_n)$. Therefore, every amart $(X_n)_{n=0,1,\ldots}$ w.r. to $(\mathcal{F}_n)$ will be amart w.r. to $(\mathcal{G}_n)$ if it is adapted to $(\mathcal{G}_n)$. Moreover, from every $(\mathcal{F}_n)$-amart (X_n) one can construct a $(\mathcal{G}_n)$-amart (Y_n) if we put $Y_n = \mathbf{E}(X_n|\mathcal{G}_n)$, $n = 0, 1, \ldots$ because of $\mathbf{E}X_\tau = \mathbf{E}Y_\tau$ for all $\tau \in T_b$.

Further, we have seen the role of "maximal inequalities" in martingale theory. It turns out, such inequality can be stated in more general setting.

Lemma 7.3 *Assume that a stochastic sequence* $(X_n)_{n=0,1,\ldots}$ *such that* $\sup_{\tau \in T_b} \mathbf{E}|X_\tau| < \infty$. *Then for* $\lambda > 0$:

$$P\left\{\sup_n |X_n| > \lambda\right\} \leq \lambda^{-1} \sup_{\tau \in T_b} \mathbf{E}|X_\tau|. \qquad (7.34)$$

7.3 Asymptotic martingales and other extensions of martingales

Proof For a fixed number N we define the set $A = \{\omega : \sup_{0 \leq n \leq N} |X_n| > \lambda\}$ and a bounded stopping time

$$\sigma(\omega) = \begin{cases} \inf(n \leq N : |X_n(\omega)| > \lambda) & if\ \omega \in A, \\ N & if\ \omega \notin A. \end{cases}$$

Then

$$\sup_{\tau \in T_b} \mathbf{E}|X_\tau| \geq \mathbf{E}|X_\sigma| \geq \lambda P(A). \tag{7.35}$$

Taking the limit in (7.35) as $N \to \infty$ we get (7.34). $\square$

Lemma 7.4 *If $(X_n)_{n=0,1,\ldots}$ and $(Y_n)_{n=0,1,\ldots}$ are L^1-bounded amarts, then $(\max(X_n, Y_n))_{n=0,1,\ldots}$ and $(\min(X_n, Y_n))_{n=0,1,\ldots}$ are amarts.*

Proof We consider only the case of $Z_n = \max(X_n, Y_n)$ due to a symmetry. First of all, we prove that the net $(\mathbf{E}Z_\tau)_{\tau \in T_b}$ is bounded. To do it take an arbitrary $\tau \in T_b$ and choose $n \geq \tau$. Define

$$\tau_X = \begin{cases} \tau, & if\ X_\tau \geq 0, \\ n, & if\ X_\tau < 0. \end{cases}$$

$$\tau_Y = \begin{cases} \tau, & if\ Y_\tau \geq 0, \\ n, & if\ Y_\tau < 0. \end{cases}$$

and find that

$$\begin{aligned} \mathbf{E}Z_\tau &\leq \mathbf{E}X_\tau I_{\{X_\tau \geq 0\}} + \mathbf{E}Y_\tau I_{\{Y_\tau \geq 0\}} \\ &= \mathbf{E}X_{\tau_X} - \mathbf{E}X_n I_{\{X_\tau < 0\}} + \mathbf{E}Y_{\tau_Y} - \mathbf{E}Y_n I_{\{Y_\tau < 0\}} \\ &\leq \sup_{\sigma \in T_b} \mathbf{E}X_\sigma + \sup_m \mathbf{E}|X_m| + \sup_{\sigma \in T_b} \mathbf{E}Y_\sigma + \sup_m \mathbf{E}|Y_m|, \end{aligned}$$

and we arrive to the conclusion.

Let us prove now that $(Z_n)_{n=0,1,\ldots}$ is an amart. For $\epsilon > 0$ we can choose $\tau_0 \in T_b$ such that

$$|\mathbf{E}X_\sigma - \mathbf{E}X_\tau| < \epsilon, \quad |\mathbf{E}Y_\sigma - \mathbf{E}Y_\tau| < \epsilon \tag{7.36}$$

for $\sigma, \tau \geq \tau_0$. Further, due to the boundedness of $(\mathbf{E}Z_\tau)_{\tau \in T_b}$ one can choose $\tau_1 \geq \tau_0$ such that if $\sigma \geq \tau_0$, then

$$\mathbf{E}Z_\sigma \leq \mathbf{E}Z_{\tau_1} + \epsilon. \tag{7.37}$$

For a bounded stopping time $\sigma \geq \tau_1$ we define $A = \{X_{\tau_1} < Y_{\tau_1}\}$ and $\sigma_1 \in T_b$ as follows

$$\sigma_1 = \begin{cases} \tau_1 & on\ A, \\ \sigma & on\ A^c. \end{cases}$$

Then we obtain

$$EX_{\tau_1} = EZ_{\tau_1} I_{A^c} + EX_{\tau_1} I_A, \qquad (7.38)$$
$$EX_{\sigma_1} = EX_\sigma I_{A^c} + EX_{\tau_1} I_A. \qquad (7.39)$$

Subtracting (7.38) from (7.37) we get with the help of (7.35) that

$$EZ_{\tau_1} I_{A^c} = EX_\sigma I_{A^c} + EX_{\tau_1} - EX_{\sigma_1} \leq EZ_\sigma I_{A^c} + \epsilon. \qquad (7.40)$$

We can write the same relations as (7.38) and (7.39) for (Y_n) :

$$EY_{\tau_1} = EZ_{\tau_1} I_{A^c} + EY_{\tau_1} I_A,$$
$$EY_{\sigma_1} = EY_\sigma I_{A^c} + EY_{\tau_1} I_A.$$

and find

$$EZ_{\tau_1} I_A = EY_\sigma I_A + EY_{\tau_1} - EY_{\sigma_1} \leq EZ_\sigma I_A + \epsilon. \qquad (7.41)$$

Combining (7.40) and (7.41), we obtain inequality

$$EZ_{\tau_1} \leq EZ_\sigma + 2\epsilon$$

which leads together with (7.36) that

$$|EZ_\sigma - EZ_{\tau_1}| \leq 2\epsilon,$$

and, hence, the net $(EZ_\tau)_{\tau \in T_b}$ is the Cauchy net. $\square$

Applying Lemma 7.3 and Lemma 7.4 we arrive to the following theorem.

Theorem 7.6 *Let* $(X_n)_{n=0,1,\ldots}$ *be amart with* $\sup_n E|X_n| < \infty$. *Then*
1) its positive (negative) parts X_n^+ (X_n^-) *and its absolute value* $|X_n|$ *are* L^1-*bounded amarts,*
2) for each $\lambda \geq 0$ *the sequence*

$$X_n^\lambda = sign(X_n)\lambda I_{|X_n| > \lambda} + X_n I_{|X_n| \leq \lambda},$$

3) $\sup_{\tau \in T_b} E|X_\tau| < \infty$ *and* $\sup_n |X_n| < \infty$ *(a.s.).*

The next statement can be considered as a version of the optional sampling theorem.

Theorem 7.7 *Let* $(X_n)_{n=0,1,\ldots}$ *be an amart for* $(\mathcal{F}_n)_{n=0,1,\ldots}$ *and let* $(\tau_m)_{m=0,1,\ldots}$ *be a non-decreasing sequence of bounded stopping times for* $(\mathcal{F}_n)_{n=0,1,\ldots}$. *Then* $(X_{\tau_m})_{m=0,1,\ldots}$ *is an amart for* $(\mathcal{F}_{\tau_m})_{m=0,1,\ldots}$.

Proof For fixed $\epsilon > 0$ we can choose N such that $|EX_\tau - EX_{\tau'}| < \epsilon$ for all bounded stopping times τ and $\tau' \geq N$. Denote $\tau_\infty = \lim_{m \to \infty} \tau_m$ (a.s.) and find that $X_{\tau_m \wedge N} \to X_{\tau_\infty \wedge N}$ (a.s.), $m \to \infty$, and

$$E \sup_m |X_{\tau_m \wedge N}| \leq E \max(|X_1,|,\ldots,|X_N|) < \infty.$$

7.3 Asymptotic martingales and other extensions of martingales

Therefore, by the dominated convergence theorem the sequence $(X_{\tau_m \wedge N})_{m=0,1,\ldots}$ is an amart. Taking a big enough M so that $|\mathbf{E}X_{\tau_\sigma \wedge N} - \mathbf{E}X_{\tau_{\sigma'} \wedge N}|$ for all bounded stopping times σ and $\sigma' \geq M$ for $(\mathcal{F}_{\tau_m})_{m=0,1,\ldots}$ we have

$$|\mathbf{E}X_{\tau_\sigma} - \mathbf{E}X_{\tau_{\sigma'}}| \leq |\mathbf{E}X_{\tau_\sigma \vee N} - \mathbf{E}X_{\tau_{\sigma'} \vee N}| + |\mathbf{E}X_{\tau_\sigma \wedge N} - \mathbf{E}X_{\tau_{\sigma'} \wedge N}| \leq \epsilon + \epsilon = 2\epsilon.$$

□

Example 7.5 An integrable stochastic sequence $(X_n)_{n=0,1,\ldots}$ is called a *quasimartingale* if

$$\sum_{n=0}^{\infty} \mathbf{E}|X_n - \mathbf{E}(X_{n+1}|\mathcal{F}_n)| < \infty. \tag{7.42}$$

Condition (7.42) is trivial in case of a martingale. In view (7.42) for given $\epsilon > 0$ one can find a big enough N such that

$$\sum_{n=N}^{\infty} \mathbf{E}|X_n - \mathbf{E}(X_{n+1}|\mathcal{F}_n)| \leq \epsilon. \tag{7.43}$$

Take $\tau \in T_b$ such that $N \leq \tau \leq M$ we have

$$|\mathbf{E}X_\tau - \mathbf{E}X_M| = \left| \sum_{m=N}^{M} \mathbf{E}(X_m - X_M) I_{\{\tau=m\}} \right|$$

$$= \left| \sum_{m=N}^{M} \sum_{n=m}^{M-1} \mathbf{E}(X_n - X_{n+1}) I_{\{\tau=m\}} \right|$$

$$= \sum_{n=N}^{M-1} \sum_{m=N}^{n} \mathbf{E}\left(|X_n - \mathbf{E}(X_{n+1}|\mathcal{F}_n)| I_{\{\tau=m\}} \right)$$

$$\leq \sum_{n=N}^{\infty} \mathbf{E}|X_n - \mathbf{E}(X_{n+1}|\mathcal{F}_n)| \leq \epsilon.$$

If $\tau_1, \tau_2 \geq N$, then choose $M \geq \tau_1 \vee \tau_2$ and get

$$|\mathbf{E}X_{\tau_1} - \mathbf{E}X_{\tau_2}| \leq |\mathbf{E}X_{\tau_1} - \mathbf{E}X_M| + |\mathbf{E}X_{\tau_2} - \mathbf{E}X_M| \leq 2\epsilon.$$

Hence, the net $(\mathbf{E}X_\tau)_{\tau \in T_b}$ is Cauchy and converges.

So, any quasimartingale is an amart.

The definition of amarts is certainly directed to provide their asymptotic behavior similar to martingales. Below we confirm these expectations.

Lemma 7.5 *Let $(X_n)_{n=0,1,\ldots}$ be a stochastic L^1-bounded sequence. Then the following are equivalent:*

(1) $(X_n)_{n=0,1,\ldots}$ converges (a.s.),
(2) $(X_n)_{n=0,1,\ldots}$ is an amart.

Proof the implication (1) $\Rightarrow$ (2) follows from the dominated convergence theorem.

To prove that (2) $\Rightarrow$ (1) we put $X^* = \limsup X_n$ and $X_* = \liminf_n X_n$. Then we can find sequences $\tau_n, \sigma_n \in T_b$ with $\tau_n \uparrow \infty$, $\sigma_n \uparrow \infty$ such that $X_{\tau_n} \to X^*$ (a.s.) and $X_{\sigma_n} \to X_*$ (a.s.). Applying again the dominated convergence theorem, we obtain that
$$\mathbf{E}(X^* - X_*) = \lim_n \mathbf{E}(X_{\tau_n} - X_{\sigma_n}) = 0.$$

It means $X^* = X_*$ (a.s.). □

Theorem 7.8 *Let $(X_n)_{n=0,1,\ldots}$ be an amart with $\sup_n \mathbf{E}|X_n| < \infty$. Then $(X_n)_{n=0,1,\ldots}$ converges (a.s.).*

Proof By Theorem 7.6, $\sup_n |X_n| < \infty$ (a.s.) and therefore $P\{\omega : \sup_n |X_n| > \lambda\}$ is arbitrary small if λ is big enough. By Theorem 7.7, X_n^λ is an amart, and by Lemma 7.5, it converges (a.s.). Taking $\lambda \uparrow \infty$ we get (a.s.)-convergence of (X_n). □

In the end of this Section we would like to mention some other extensions of the notion of martingales.

Definition 7.4 A stochastic sequence $(X_n)_{n=0,1,\ldots}$, $X_0 = 0$, is a **local martingale** if there exists a sequence of stopping times $(\tau_m)_{m=1,2,\ldots}$ increasing to $+\infty$ such that $(X_{\tau_m \wedge n})_{n=0,1,\ldots}$ is a martingale for each $m = 1, 2, \ldots$. The sequence (τ_m) is called a *localizing* sequence.

Definition 7.5 A stochastic sequence $(X_n)_{n=0,1,\ldots}$, $X_0 = 0$, is a *generalized martingale* if (a.s.)
$$\{\omega : \mathbf{E}(X_n^+|\mathcal{F}_{n-1}) < \infty\} \cup \{\omega : \mathbf{E}(X_n^-|\mathcal{F}_{n-1}) < \infty\} = \Omega,$$

and for $n = 1, 2, \ldots$
$$\mathbf{E}(X_n|\mathcal{F}_{n-1}) = X_{n-1} \quad (a.s.)$$

Definition 7.6 A stochastic sequence $(X_n)_{n=0,1,\ldots}$, $X_0 = 0$, is a *martingale transform* (stochastic integral) if it admits the following representation
$$X_n = \sum_{m=0}^{n} H_m \Delta M_m,$$

where $(M_n)_{n=0,1,\ldots}$, $M_0 = 0$, is a martingale, $(H_n)_{n=0,1,\ldots}$ is predictable, $H_0 = \Delta M_0 = M_0 = 0$.

All these generalizations are directed to relax the condition of integrability of stochastic sequences. All these classes of stochastic sequences are coincide. We omit the proof of this statement here.

Chapter 8
Elements of classical theory of stochastic processes

Abstract This chapter contains a general notion of random processes with continuous time. It is given in context of the Kolmogorov consistency theorem. The notion of a Wiener process with variety of its properties are also presented here. Its existence is stated by two ways: with the help of the Kolmogorov theorem as well as with the help of orthogonal functional systems. Besides the Wiener process as a basic process for many others, the Poisson process is also considered here. Stochastic integration with respect to Wiener process is developed for a class of progressively measurable functions. It leads to the Ito processes, the Ito formula, the Girsanov theorem and representation of martingales (see [5], [6], [14], [17], [21], [35], [41], and [44]).

8.1 Stochastic processes: definitions, properties and classical examples

In classical theory the notion of a stochastic process is associated with a family of random variables (X_t) on a probability space $(\Omega, \mathcal{F}, P)$ with values in space $\mathbb{R}^d, d \geq 1$, where parameter $t \in [0, \infty) = \mathbb{R}_+$. Usually for simplicity we put $d = 1$. It is supposed that for time parameter $t \in \mathbb{R}_+$ we have a random variable $X_t(\omega), \omega \in \Omega$. On the other hand, one can fix $\omega \in \Omega$ and get a function $X_{\cdot}(\omega): \mathbb{R}_+ \to \mathbb{R}^d$, which is called a *trajectory*. This point of view opens up the possibility of studying stochastic processes that have been exhaustively determined by their distributions as probability measures on the functional space $(\mathbb{R}^{[0,\infty)}, \mathcal{B}^{[0,\infty)})$. Usually a family of finite dimensional distributions $P_{t_1,\ldots,t_n}(B_1, \ldots, B_n), 0 \leq t < t_1 < t_2 < \ldots < t_n < \infty$, $B_i \in \mathcal{B}(\mathbb{R}^d)$, $i = 1, \ldots, n$, is exploited as the basic probabilistic characteristic of a stochastic process. A natural question arises in this regard. Is there a stochastic process $(X_t(\omega))$ such that $P(X_{t_1} \in B_1, \ldots, X_{t_n} \in B_n) = P_{t_1,\ldots,t_n}(B_1, \ldots, B_n)$. As it was noted in Chapter 1, this question is solved with the help of the Kolmogorov theorem if for the system $(P_{t_1,\ldots,t_n})$ the following *consistency conditions* are fulfilled:

1. $P_{t_1,\ldots,t_n}(B_1, \ldots, B_{i-1}, \cdot, B_{i+1}, \ldots, B_n)$, $i = 1, 2, \ldots, n$, is a probability measure on $(\mathbb{R}^d, \mathcal{B}(\mathbb{R}^d))$;

© The Author(s), under exclusive license to Springer Nature Switzerland AG 2023
A. Melnikov, *A Course of Stochastic Analysis*, CMS/CAIMS Books in Mathematics 6,
https://doi.org/10.1007/978-3-031-25326-3_8

2. $P_{t_1,\ldots,t_n}(B_1,\ldots,B_n) = P_{t_{i_1},\ldots,t_{i_n}}(B_{i_1},\ldots,B_{i_n})$, where $(i_1,\ldots,i_n)$ is arbitrary permutation of numbers $(1,2,\ldots,n)$;

3. $P_{t_1,\ldots,t_{n-1},t_n}(B_1,\ldots,B_{n-1},\mathbb{R}^d) = P_{t_1,\ldots,t_{n-1}}(B_1,\ldots,B_{n-1})$. In such a case there exists a probability measure P_X in the functional space $(\mathbb{R}^{[0,\infty)}, \mathcal{B}^{[0,\infty)})$ and the process $(X_t(\omega))$ is constructed as $X_t(\omega) = \omega_t$, where ω_t is the value of function $\omega. \in \mathbb{R}^{[0,\infty)}$ at time t. Measure P_X will be a distribution of $(X_t(\omega))$ which has $P_{t_1,\ldots,t_n}$ as a system of finite dimensional distributions of this process.

After these necessry explanations we introduce the notion of a *Wiener process* or *Brownian motion*.

Definition 8.1 A stochastic process $(W_t)_{t \geq 0}$ such that
1) $W_0 = 0$ $(a.s.)$,
2) $W.(\omega) \in C[0,\infty)$ for almost all $\omega \in \Omega$,
3) $P(W_{t_1} \in B_1, \ldots, W_{t_n} \in B_n) = \int_{B_1} \ldots \int_{B_n} p(t, 0, x_1) p(t_2 - t_1, x_1, x_2) \ldots p(t_n - t_{n-1}, x_{n-1}, x_n) dx_1 \ldots dx_n$,

for arbitrary $0 < t_1 < t_2 < \ldots t_n$, $B_1, \ldots, B_n \in \mathcal{B}(\mathbb{R})$, where $p(t,x,y) = \frac{1}{\sqrt{2\pi t}} \exp\left(-\frac{(x-y)^2}{2t}\right)$ is called a Wiener process.

It follows directly from the definition that $W_t \sim N(0,t)$ for every fixed $t \in \mathbb{R}_+$.

Let us formulate the first *properties* of the process $(W_t)_{t \geq 0}$:
(1) $\mathbf{E} W_t W_s = s \wedge t$ for arbitrary $s, t \in \mathbb{R}_+$;
(2) $\mathbf{E}(W_t - W_s)^2 = |t - s|$;
(3) $\mathbf{E}(W_t - W_s)^4 = 3(t-s)^2$.

We only prove (2), leaving (1) and (3) as problems.

Problem 8.1 Prove properties (1) and (3) of $(W_t)_{t \geq 0}$.

Assume that $s < t$ and find that

$$\mathbf{E} W_s W_t = \int_{-\infty}^{\infty} \int_{-\infty}^{\infty} xy p(s, 0, x) p(t-s, x, y) dx dy$$

$$= \int_{-\infty}^{\infty} x p(s, 0, x) \left(\int_{-\infty}^{\infty} y p(t-s, x, y) dy \right) dx$$

$$= \int_{-\infty}^{\infty} x p(s, 0, x) \left(\int_{-\infty}^{\infty} (x+u) p(t-s, x, x+u) du \right) dx$$

$$= \int_{-\infty}^{\infty} x p(s, 0, x) \left(\int_{-\infty}^{\infty} (x+u) p(t-s, 0, u) du \right) dx$$

$$= \int_{-\infty}^{\infty} x p(s, 0, x) \left(x \int_{-\infty}^{\infty} p(t-s, 0, u) du + u \int_{-\infty}^{\infty} p(t-s, 0, u) du \right) dx$$

$$= \int_{-\infty}^{\infty} x p(s, 0, x) (x + 0) dx$$

$$= \int_{-\infty}^{\infty} x^2 p(s, 0, x) dx = s.$$

Now we calculate for $s < t$ and $B \in \mathcal{B}(\mathbb{R})$

8.1 Stochastic processes: definitions, properties and classical examples

$$P(W_t - W_s \in B) = \int_{\{(x,y):y-x\in B\}} p(s,0,x)p(t-s,x,y)dxdy$$

$$= \int_{-\infty}^{\infty} p(s,0,x)\left(\int_{\{y:y-x\in B\}} p(t-s,x,y)dy\right)dx$$

$$= \int_{-\infty}^{\infty} p(s,0,x)\left(\int_B p(t-s,x,x+u)du\right)dx$$

$$= \int_{-\infty}^{\infty} p(s,0,x)\left(\int_B p(t-s,0,u)du\right)dx$$

$$= \int_{-\infty}^{\infty} p(s,0,x)dx \int_B p(t-s,0,u)du = \int_B p(t-s,0,u)du$$

and we obtain that $W_t - W_s \sim N(0, t-s)$.

As a result, we arrive to the fourth property of W_t:

(4) For $0 =_t 0 < t_1 < \ldots < t_n$ the increments $W_{t_1} - W_{t_0}, \ldots, W_{t_n} - W_{t_{n-1}}$ of a Wiener process are *independent* random variables.

Proof Due to a "normality" of increments we need to prove that they are *uncorrelated*. Let us take $r < s < t < u$ and find that

$$\mathbf{E}(W_u - W_t) = \mathbf{E}W_u W_s - \mathbf{E}W_t W_s - \mathbf{E}W_u W_r + \mathbf{E}W_t W_r = s - s - r + r = 0.$$

$\square$

To go further, we introduce filtration, generated by a Wiener process $\mathcal{F}_t = \sigma(W_s, s \leq t)$. We assume that $\mathcal{F}_t$ is *complete*, i.e. it contains all sets of P–measure zero.

Definition 8.2 A process $(M_t)_{t\geq 0}$ satisfying conditions
  1) $M_t - \mathcal{F}_t$-measurable,
  2) M_t is integrable,
  3) $\mathbf{E}(M_t|\mathcal{F}_s) = M_s$ (a.s.) for all $s < t$,
is called a *martingale*.

After this definition we formulate the *martingale* property of a Wiener process. If in (3) the equality is replaced by the inequality $\geq$ ($\leq$), then (M_t) is a submartingale (supermartingale).

(5) Processes $(W_t)_{t\geq 0}$ and $(W_t^2 - t)_{t\geq 0}$ are martingales w.r.to $(\mathcal{F})_{t\geq 0}$.

Remark 8.1 In fact, the inverse statement to (5) is well-known as *Levy's characterization of a Wiener process*.

(6) The Doob maximal inequality for a Wiener process $(W_t)_{t\geq 0}$ has the following form: for $t > 0$,

$$\mathbf{E} \max_{s\leq t} |W_s|^2 \leq 4\mathbf{E}|W_t|^2. \tag{8.1}$$

To prove it we consider a subdivision of the interval $[0, t]$ as of the interval $[0, t]$ as $t_k^{(n)} = \frac{kt}{2^n}, 0 \leq k \leq 2^n$, and define a square integrable martingale with discrete time

$$M_k = M_k^n = W_{\frac{kt}{2^n}}, \mathcal{F}_k^n = \mathcal{F}_{\frac{kt}{2^n}}, 0 \le k \le 2^n.$$

For $(|M_k|)$ we have, using its submartingale property that

$$\mathbf{E}(\max_k |M_k|)^2 = 2\int_0^\infty y P(\max_k |M_k| > y) dy$$

$$\le 2\int_0^\infty \mathbf{E}|M_{2^n}| I_{(\max_k |M_k| \ge y)} dy$$

$$= 2\int_0^\infty \left(\int_{(\max_k |M_k| \ge y)} |M_{2^n}| dP\right) dy$$

$$= 2\int_\Omega |M_{2^n}^k| \left(\int_0^{\max_k |M_k|} dy\right) dP$$

$$= 2\int_\Omega |M_{2^n}| \max_k |M_k| dP = 2\mathbf{E}|M_{2^n}| \max_k |M_k|$$

$$\le 2(\mathbf{E}(M_{2^n})^2)^{1/2} \left(\mathbf{E}(\max_k |M_k|)^2\right)^{1/2}.$$

As a result, we obtain that

$$(\mathbf{E}(\max_k |M_k|)^2)^{1/2} \le 2(\mathbf{E}(M_{2^n})^2)^{1/2},$$

and, hence,

$$\mathbf{E}\max_k |M_k^n|^2 \le \mathbf{E}(\max_k |M_k^n|)^2 \le 4\mathbf{E}(M_{2^n})^2 = 4\mathbf{E}|W_t|^2. \tag{8.2}$$

Taking the limit as $n \to \infty$ in (8.2) and using the continuity of trajectories of a Wiener process we arrive to inequality (8.1).

(7) For any positive constant c the process $W_t^c = c^{-1} W_{c^2 t}$ is a Wiener process. This property of a Wiener process is called *self-similarity*.

To prove the self-similarity of a Wiener process one can use the Levy characterization of this process.

(8) *Existence* of a Wiener process follows from the consistency theorem of Kolmogorov (see Chapter 1). However, such a reference is incomplete because this theorem guarantees existence of this process only in space $(\mathbb{R}^{[0,\infty)}, \mathcal{B}^{[0,\infty)})$. To make the proof complete one can use another *theorem of Kolmogorov:*

Suppose $(X_t)_{t \ge 0}$ is a stochastic process satisfying the following condition

$$\mathbf{E}|X_t - X_s|^\alpha \le C|t-s|^{1+\beta} \text{ for all } t, s \ge 0,$$

and for some $\alpha > 0, \beta > 0, 0 < C < \infty$. Then the process (X_t) admits a continuous modification i.e. there exists a continuous process (Y_t) such that $P(X_t = Y_t) = 1$ for all $t \in \mathbb{R}_+$.

Property (3) of a Wiener process (W_t) means that this theorem can be applied, and therefore, the process (W_t) can be identified with its continuous modification.

8.1 Stochastic processes: definitions, properties and classical examples

(9) Now we give some other characterizations of trajectories of $(W_t)_{t\geq 0}$. The continuity property brings an element of a regularity of trajectories. But this description of sample paths properties is clearly incomplete as it is shown below.

Let us analyze the *first* question of *differentiability* of trajectories of $(W_t)_{t\geq 0}$. We take $t \geq 0$ and consider the ratio $R_\Delta^W(t) = \frac{W_{t+\Delta t} - W_t}{\Delta t}$, where $\Delta t \to 0$. One can easily observe that

$$\mathbf{E} R_\Delta^W(T) = \frac{\mathbf{E} W_{t+\Delta t} - \mathbf{E} W_t}{\Delta t} = 0,$$

and

$$\mathbf{Var} R_\Delta^W(t) = (\Delta t)^{-2} \mathbf{Var}(W_{t+\Delta t} - W_t) = (\Delta t)^{-2} \Delta t = (\Delta t)^{-1} \to \infty \text{ as } \Delta t \to 0.$$

It shows that $W_\cdot$ is *not differentiable* in the L^2–*sense*. Moreover, one can prove that almost all trajectories of W are non-differentiable.

Let us analyze the second question of how *vary* the trajectories of $(W_t)_{t\geq 0}$ are. This property is characterized with the help of the notion of *p-variation* of given function $f : [0, T] \to \mathbb{R}, p \geq 1$:

We say that

$$\limsup_{\Delta t \to 0} \sum_{i=1}^n |f(t_i) - f(t_{i-1})|^p, 0 = t_0 < t_1 < \ldots < t_n = T,$$

is a p–variation of f on $[0, T]$.

Usually, the first and second variations are most important in this regard.

For the second variation of $(W_t)_{t\geq 0}$. We have the following observations. In view of independent increments we obtain for $0 = t_0 < t_1 < \ldots < t_n = T$ that $\mathbf{E} \sum_{i=1}^n (W_{t_i} - W_{t_{i-1}})^2 = \sum_{i=1}^n \mathbf{E}(W_{t_i} - W_{t_{i-1}})^2 = \sum_{i=1}^n \mathbf{Var}(W_{t_i} - W_{t_{i-1}}) = \sum_{i=1}^n (t_i - t_{i-1}) = T$.

Further, with the help of (3):

$$\mathbf{Var} \sum_{i=1}^n (W_{t_i} - W_{t_{i-1}})^2 = \sum_{i=1}^n \mathbf{Var}(W_{t_i} - W_{t_{i-1}})^2$$

$$= \sum_{i=1}^n \left[\mathbf{E}(W_{t_i} - W_{t_{i-1}})^4 - \left(\mathbf{E}(W_{t_i} - W_{t_{i-1}})^2\right)^2 \right]$$

$$= \sum_{i=1}^n \left[3(t_i - t_{i-1})^2 - (t_i - t_{i-1})^2 \right]$$

$$= 2 \sum_{i=1}^n (t_i - t_{i-1})^2.$$

Therefore,

$$\mathbf{E}\left[\sum_{i=1}^{n}(W_{t_i}-W_{t_{i-1}})^2 - T\right]^2 = \mathbf{Var}\left(\sum_{i=1}^{n}(W_{t_i}-W_{t_{i-1}})^2\right)$$

$$= 2\sum_{i=1}^{n}(t_i - t_{i-1})^2 \leq 2T \max_{0<i\leq n}(t_i - t_{i-1}) \to 0,$$

and we conclude that the second variation of (W_t) on $[0,T]$ converges to the *length* of this interval in L^2 *sense*, and hence, in the *sense of convergence in probability*. To investigate the *first variation* of (W_t) we consider the probability $P(\omega : \sum_{i=1}^{n} |W_{t_i} - W_{t_{i-1}}| > N)$. First of all, we note that $\mathbf{E}\sum_{i=1}^{n}|W_{t_i} - W_{t_{i-1}}| = \sqrt{\frac{2}{\pi}}\sum_{i=1}^{n}\sqrt{t_i - t_{i-1}} \to \infty$ as $\max(t_i - t_{i-1}) \to 0$, and

$$\mathbf{Var}\sum_{i=1}^{n}|W_{t_i} - W_{t_{i-1}}| = \sum_{i=1}^{n}(1 - \frac{2}{\pi})(t_i - t_{i-1}) = (1 - \frac{2}{\pi})T.$$

Further, for $\mathbf{E}\sum_{i=1}^{n}|W_{t_i} - W_{t_{i-1}}| > n$ with the help of the Chebyshev inequality we get

$$P(\omega : \sum_{i=1}^{n}|W_{t_i} - W_{t_{i-1}}| \leq N) \leq P(\omega : \sum_{i=1}^{n}|W_{t_i} - W_{t_{i-1}}| - \mathbf{E}\sum_{i=1}^{n}|W_{t_i} - W_{t_{i-1}}| \geq \mathbf{E}\sum_{i=1}^{n}|W_{t_i} - W_{t_{i-1}}| - N)$$

$$\leq \mathbf{Var}(\sum_{i=1}^{n}|W_{t_i} - W_{t_{i-1}}|)/(\mathbf{E}\sum_{i=1}^{n}|W_{t_i} - W_{t_{i-1}}| - N)^2 \to 0,$$

as $\max(t_i - t_{i-1}) \to 0$.

It means that the first variation of (W_t) *converges in probability to* ∞.

Remark 8.2 It is possible to prove that all mentioned results are true in the sense of (a.s.)-convergence.

Another process that plays an important role in both theory and applications is a Poisson process.

Definition 8.3 The process $(N_t)_{t\geq 0}$ is called a *Poisson process* with parameter $\lambda > 0$, if the following conditions are satisfied:

1) $N_0 = 0$. (a.s.),

2) its increments $N_{t_1} - N_{t_0}, \ldots, N_{t_n} - N_{t_{n-1}}$ are independent random variables for any sundivision $0 \leq t_0 < t_1 < \ldots < t_n < \infty$,

3) $N_t - N_s$ is a random variable that has a Poisson distribution with parameter $\lambda(t - s)$:

$$P(\omega : N_t - N_s = i) = e^{-\lambda(t-s)}\frac{(\lambda(t-s))^i}{i!}, 0 \leq s \leq t < \infty, i = 0, 1, \ldots.$$

Let us present some properties of sample paths of this process.

8.1 Stochastic processes: definitions, properties and classical examples 87

First of all, we note that its almost all trajectories are *non-decreasing*, because for $s \leq t$:

$$P(\omega : N_t - N_s \geq 0) = \sum_{i=0}^{\infty} P(\omega : N_t - N_s = i) = \sum_{i=0}^{\infty} e^{-\lambda(t-s)} \frac{(\lambda(t-s))^i}{i!}$$

$$= e^{-\lambda(t-s)} \sum_{i=0}^{\infty} \frac{(\lambda(t-s))^i}{i!} = e^{-\lambda(t-s)} e^{\lambda(t-s)} = 1.$$

The process $(N_t)_{t \geq 0}$ is *stochastically continuous*: in the sense $N_t \to N_s$ in probability as $t \to s$. This property is an obvious consequence of the application of the Chebyshev inequality. It is an interesting effect when a jumping process satisfies a continuity property. Moreover, one can say something about a *differentiability* of trajectories of $(N_t)_{t \geq 0}$ in the sense of convergence in probability, i.e.

$$\frac{N_{t+\Delta t} - N_t}{\Delta t} \xrightarrow[\Delta t \to 0]{p} 0. \qquad (8.3)$$

Problem 8.2 Prove relation (8.3).

Further, one can derive from $P(\omega : N_s \leq N_t) = 1$ for all $s \leq t$ that there is a *right-continuous modification* of (N_t), almost all trajectories of which are non-decreasing integer-valued functions with unit jumps.

Let us note a martingale property of $(N_t)_{t \geq 0}$. To do this we introduce a natural filtration $(\mathcal{F}_t)_{t \geq 0}$, generated by $(N_t)_{t \geq 0}$ and completed by sets of P-measure zero. We note that

$$\mathbf{E} N_t = \sum_{i=0}^{\infty} i P(\omega : N_t = i) = \sum_{i=0}^{\infty} i e^{-\lambda t} \frac{(\lambda t)^i}{i!}$$

$$= (\lambda t) e^{-\lambda t} \sum_{i=1}^{\infty} \frac{(\lambda t)^{i-1}}{(i-1)!} = (\lambda t) e^{-\lambda t} e^{\lambda t} = \lambda t.$$

Let us define a new process $M_t = N_t - \lambda t$ and find that for $s \leq t$ due to independence $(N_t - N_s)$ of $\mathcal{F}_s$:

$$\mathbf{E}(M_t | \mathcal{F}_s) = \mathbf{E}(N_t - \lambda t | \mathcal{F}_s) = \mathbf{E}(N_t - N_s + N_s - \lambda t | \mathcal{F}_s) \mathbf{E}(N_t - N_s | \mathcal{F}_s) + N_s - \lambda t$$

$$= \mathbf{E}(N_t - N_s) + N_s - \lambda t = \lambda(t-s) + N_s - \lambda t = N_s - \lambda s = M_s.$$

Hence, $(M_t)_{t \geq 0}$ is a *martingale* w.r.to $(\mathcal{F}_t)_{t \geq 0}$.

We have seen that both processes $(W_t)_{t \geq 0}$ and (N_t) are stochastically continuous processes with independent increments. The following example shows why a consideration of *stochastic processes with independent values* is not productive.

Example 8.1 Let $(X_t)_{t \geq 0}$ be a family of independent random variables with the same density $f = f(x) \geq 0$, $\int_{-\infty}^{\infty} f(x)dx = 1$. Then for a fixed $s \geq 0$, and $t \neq s$, $\epsilon > 0$ we have that

$$P(\omega : |X_t - X_s| \geq \epsilon) = \int\int_{|x-y|\geq\epsilon} f(x)f(y)dxdy. \tag{8.4}$$

The integral in the right hand side of equality (8.4) converges (as $\epsilon \to 0$) to

$$\int\int_{x\neq y} f(x)f(y)dxdy = \int_{-\infty}^{\infty}\int_{-\infty}^{\infty} f(x)f(y)dxdy. \tag{8.5}$$

The relation (8.5) shows that for some $\epsilon > 0$ the probability in the left hand side of (8.4) does not converge to zero as $t \to s$. It means the process with independent values is not stochastically continuous even.

The existence of a Wiener process was derived from the Kolmogorov consistency theorem. Such a derivation does not look constructive. That is why we want to demonstrate *a direct way of construction* of a Wiener process, based on orthogonal systems of functions of Haar and Schauder, and sequences of independent normal random variables.

Let us define a system of functions $(H_k(t))_{k=1,2,\ldots}, t \in [0,1]$, called the *Haar system*:

$$H_1(t), H_2(t) = I_{[0,2^{-1}]}(t) - I_{(2^{-1},1]}(t), \ldots,$$
$$H_k(t) = 2^{n/2}\left(I_{(a_{n,k}, a_{n,k}+2^{-n-1}]} - I_{(a_{n,k}+2^{-n-1}, a_{n,k}+2^{-n}]}(t)\right)$$

where $w^n < k \leq 2^{n+1}$, $a_{n,k} = 2^{-n}(k - 2^n - 1)$, $n = 1, 2, \ldots$.

In the space $L^2([0,1], \mathcal{B}(0,1), dt)$ with the scalar product $\langle f, g \rangle = \int_0^1 f_s g_s ds$, $f, g \in L^2$, the system $(H_k(t))_{k=1,2,\ldots}$ is complete and orthogonal. Hence,

$$f = \sum_{k=1}^{\infty}\langle f, H_k\rangle H_k, \quad g = \sum_{k=1}^{\infty}\langle g, H_k\rangle H_k, \quad \langle f, g\rangle = \sum_{k=1}^{\infty}\langle f, H_k\rangle\langle g, H_k\rangle.$$

Using the system $(H_k(t))_{k=1,2,\ldots}$ one can construct the *Schauder system* $(S_k(t))_{k=1,2,\ldots}$ as follows

$$S_k(t) = \int_0^t H_k(y)dy = \langle I_{[0,t]}, H_k\rangle, \quad t \in [0,1].$$

Lemma 8.1 *If a sequence of real numbers $a_k = O(k^\epsilon)$, $k \to \infty$, for some $\epsilon \in (0, 1/2)$, then the series $\sum_{k=1}^{\infty} a_k S_k(t)$ converges uniformly on $[0,1]$, and, hence, it is a continuous function.*

Proof Denote $R_m = \sup_t \sum_{k>2^m} |a_k| S_k(t)$, $m = 1, 2, \ldots$, and note that $|a_k| \leq ck^\epsilon$ for all $k \geq 1$ and some constant $c > 0$. Therefore, for all t and $n = 1, 2, \ldots$ we obtain that

$$\sum_{2^n < k < 2^{n+1}} |a_k| S_k(t) \leq c 2^{(n+1)\epsilon} \sum_{2^n < k < 2^{n+1}} S_k(t)$$
$$\leq c 2^{(n+1)\epsilon} 2^{-n/2-1}$$
$$\leq c 2^{\epsilon - n(1/2-\epsilon)}. \tag{8.6}$$

8.1 Stochastic processes: definitions, properties and classical examples

It follows from here that $R_m \leq c2^\epsilon \sum_{n \geq m} 2^{-n(1/2-\epsilon)} \to 0$ as $m \to \infty$. □

Lemma 8.2 *Let $(\xi_k)_{k=1,2,\ldots}$ be a sequence of standard normal random variable on a complete probability space $(\Omega, \mathcal{F}, P)$. Then for every constant $c > \sqrt{2}$ and almost all $\omega \in \Omega$ there exists a number $N_0 = N_0(\omega, c)$ such that $|\xi_k(\omega)| < c\sqrt{\ln k}$ for all $k \geq N_0$.*

Proof For a standard normal random variable $\xi \sim N(0,1)$ and $x > 0$ we have that

$$\begin{aligned} P(\omega : \xi \geq x) &= (2\pi)^{-1/2} \int_x^\infty e^{-y^2/2} dy \\ &= (2\pi)^{-1/2} \int_x^\infty \left(-\frac{1}{y}\right) de^{-y^2/2} \\ &= (2\pi)^{-1/2} \left(x^{-1} e^{-x^2/2} - \int_x^\infty y^{-2} e^{-y^2/2} dy\right) \\ &\leq x^{-1} (2\pi)^{-1/2} (x^{-1} e^{-x^2/2}) \end{aligned} \tag{8.7}$$

and, hence,

$$P(\omega : |\xi| \geq x) \leq x^{-1}(2\pi)^{1/2} e^{-x^2/2}.$$

Applying the above inequality we can estimate the series below as follows

$$\sum_{k \geq 2} P\left(\omega : |\xi_k| \geq c(\ln k)^{1/2}\right) \leq c^{-1}(2/\pi)^{1/2} \sum_{k \geq 2} k^{-c^2/2} (\ln k)^{-1/2} < \infty,$$

and the statement of Lemma 8.2 follows from the Borel-Cantelli lemma. □

Theorem 8.1 *Let $(\xi_k)_{k=1,2,\ldots}$ be a sequence of independent standard normal random variables $(\Omega, \mathcal{F}, P)$. Define a stochastic process $W_t = W(t, \omega) = \sum_{k=1}^\infty \xi_k(\omega) S_k(t)$, $t \in [0,1]$, $\omega \in \Omega$. Then $(W_t)_{t \in [0,1]}$ is a Wiener process.*

Proof First of all, we note that (W_t) has continuous trjectories due to Lemma 8.1 and 8.2.

Denote $Z_n(t) = \sum_{k=1}^n \xi_k S_k(t)$, $t \in [0,1]$, and find that

$$\begin{aligned} \mathbf{E}|Z_{n+m}(t) - Z_n(t)|^2 &= \mathbf{E} \left| \sum_{k=n+1}^{n+m} \xi_k S_k(t) \right|^2 \\ &= \sum_{k=n+1}^{n+m} \sum_{l=n+1}^{n+m} S_k(t) S_l(t) \mathbf{E}\xi_k \xi_l \\ &= \sum_{k=n+1}^{n+m} S_k^2(t). \end{aligned} \tag{8.8}$$

Further, $\sum_{k=1}^\infty S_k^2(t) \leq \sum_{k=1}^\infty \left(2^{-k/2-1}\right)^2 < \infty$, and, hence, due to a completeness of $L^2(\Omega, \mathcal{F}, P)$ we arrive to conclusion that there exists a limit $Z(t)$ of $Z_n(t)$ in this space such that

$$E|Z_n(t) - Z(t)|^2 \to 0, \ n \to \infty, \ t \in [0, 1].$$

This limit $Z(t)$ coincides with W_t up to equivalence and $Z_n(t) \xrightarrow{t} W_t$ in L^2-sense. Now due to $\mathbf{E}Z_n(t) = 0$ we obtain that

$$|\mathbf{E}W_t| = |\mathbf{E}W_t - \mathbf{E}Z_n(t)| \leq \left(\mathbf{E}|W_t - Z_n(t)|^2\right)^{1/2} \xrightarrow[n \to \infty]{} 0,$$

which means that $\mathbf{E}W_t = 0$.

Using a continuity property of the scalar product we find that

$$\langle Z_n(s), Z_n(t) \rangle_{L^2(\Omega)} \to \langle W_s, W_t \rangle_{L^2(\Omega)} = \mathbf{E}W_s W_t = \mathbf{Cov}(W_s, W_t), \ s, t \in [0, 1].$$

On the other hand, we can use the independence of $(\xi_k)_{k=1,2,\ldots}$ and $\mathbf{E}\xi_k = 0$, $\mathbf{E}\xi_k^2 = 1$, and observe that

$$\langle Z_n(s), Z_n(t) \rangle_{L^2(\Omega)} = \mathbf{E}Z_n(s)Z_n(t) = \sum_{k=1}^{n} S_k(s)S_k(t)\mathbf{E}\xi_k^2 \to \sum_{k=1}^{\infty} S_k(t)S_k(t), \ n \to \infty.$$

Applying the Parseval equality for scalar products we finally derive that

$$\sum_{k=1}^{\infty} S_k(s)S_k(t) = \sum_{k=1}^{\infty} \langle H_k, I_{[0,s]} \rangle \langle H_k, I_{[0,t]} \rangle = \langle I_{[0,s]}, I_{[0,t]} \rangle = \min(s, t) = \mathbf{Cov}(W_s, W_t).$$

Let us prove that $(W_t)_{t \in [0,1]}$ is a Gaussian process. Take $\mu = (\mu_1, \ldots, \mu_n) \in \mathbb{R}^n$, $t_1, \ldots, t_n \in [0, 1]$ and define $Y = \sum_{m=1}^{n} \mu_m W_{t_m}$. We have now that

$$Y = \sum_{m=1}^{n} \mu_m \sum_{k=1}^{\infty} \xi_k S_k(t_m) = \sum_{k=1}^{\infty} b_k \xi_k,$$

where $b_k = b_k(\mu_1 \ldots \mu_n, t_1, \ldots, t_n) = \sum_{m=1}^{n} \mu_m S_k(t_m)$. Consider $Y_N = \sum_{k=1}^{N} b_k \xi_k \sim N(0, \sigma_N^2)$, $\sigma_N^2 = \sum_{k=1}^{N} b_k^2$ and find that

$$\mathbf{E}Y_N^2 = \sigma_N^2 \to \mathbf{E}Y^2 = \sigma^2, \ N \to \infty,$$

in the sense of $L^2(\Omega)$. Therefore, we have also the convergence in distribution $Y_N \xrightarrow{d} Y$, $N \to \infty$. Hence, we have that the convergence of

$$\phi_Y(\lambda) = \exp(-\sigma_N^2 \lambda^2 / 2) \to \exp(-\sigma^2 \lambda^2 / 2) = \phi_Y(\lambda), \ \lambda \in \mathbb{R},$$

i.e. $Y \sim N(0, s\sigma^2)$.

Finally, we can extend this construction from the unit time interval to the whole $[0, \infty)$ as follows

$$W(t, \omega) = \begin{cases} W_t = W(t, \omega) = W_1(t, \omega), & t \in [0, 1), \\ \sum_{j=1}^{k} W_j(1, \omega) + W_{k+1}(t - k, \omega), & t \in [k, k+1), \ k = 1, 2, \ldots, \end{cases}$$

where $W_n = (W_n(t))_{t \in [0,1]}$ be a sequence of independent Wiener processes. □

8.2 Stochastic integrals with respect to a Wiener process

Let $(\Omega, \mathcal{F}, P)$ be a complete probability space, $(W_t)_{t \geq 0}$ be a Wiener process and $(\mathcal{F}_t)_{t \geq 0}$ be a complete filtration generated by this process.

We describe here a scheme of construction of a stochastic integral

$$I(f) = \int_0^\infty f_s(\omega) dW_s \tag{8.9}$$

for any $\mathcal{F}_t$-adapted and $\mathcal{B}(0, \infty) \times \mathcal{F}$-measurable random function/process $f_t(\omega) = f(t, \omega)$ integrable in square with respect to $dt \times dP$:

$$\mathbf{E} \int_0^\infty f_s^2 ds < \infty. \tag{8.10}$$

The class of these random functions is denoted S^2, and the stochastic integral (8.9) must be a *linear* and *isometric operator* from $L^2(\mathbb{R}_+ \times \Omega, \mathcal{B}(\mathbb{R}_+) \times \mathcal{F}, dt \times dP)$ to $L^2(\Omega, \mathcal{F}, dP)$. We start its construction from the class of *step functions* $f \in S_0^2 \subseteq S^2$. For each function f there exists a subdivision $0 = t_0 \leq t_1 \leq \ldots \leq t_n < \infty$ such that f_{t_i}-$\mathcal{F}_{t_i}$- measurable and square-integrable, and $f_t = f_{t_i}$ for $t \in [t_i, t_{i+1})$, $i \leq n$, whereas $f_t = 0$ for $t \geq t_n$. Hence, the weighted sum

$$\sum_{i=0}^{n-1} f_{t_i} I_{[t_i, t_{i+1})}(t) = f_t(\omega). \tag{8.11}$$

For function $f_t(\omega)$, defined by (8.11) we introduce a stochastic integral $I(f)$ as follows

$$I(f) = \int_0^\infty f_s dW_s = \sum_{i=0}^{n-1} f_{t_i}(W_{t_{i+1}} - W_{t_i}). \tag{8.12}$$

It is almost obvious that the definition of $I(f)$ in (8.12) does not depend on changes of subdivisions (t_i), and it is a linear mapping from S_0^2 to $L^2(\Omega, \mathcal{F}, P)$, i.e.

$$I(\alpha f + \beta g) = \alpha I(f) + \beta I(g), \tag{8.13}$$

where $\alpha, \beta \in \mathbb{R}, f, g \in S_0^2$.

Using properties (1)-(2) and (4) of a Wiener process $(W_t)_{t \geq 0}$ and properties of conditional expected values, we obtain that

$$\mathbf{E} I^2(f) = \mathbf{E}\left[\sum_{i=0}^{n-1} f_{t_i}^2 (W_{t_{i+1}} - W_{t_i})^2 + 2\sum_{j<i} f_{t_j}(W_{t_{j+1}} - W_{t_j})f_{t_i}(W_{t_{i+1}} - W_{t_i})\right]$$

$$= \mathbf{E}\left[\sum_{i=0}^{n-1} f_{t_i}^2 \mathbf{E}\left((W_{t_{i+1}} - W_{t_i})^2 | \mathcal{F}_{t_i}\right) + 2\sum_{j<i} f_{t_j}(W_{t_{j+1}} - W_{t_j})f_{t_i} \mathbf{E}\left((W_{t_{i+1}} - W_{t_i}) | \mathcal{F}_{t_i}\right)\right]$$

$$= \sum_{i=0}^{n-1} \mathbf{E} f_{t_i}^2 (t_{i+1} - t_i) = \mathbf{E}\int_0^\infty f_s^2 \, ds. \tag{8.14}$$

The *isometry* relation (8.14) means that the norms $\|I(f)\|_{L^2(\Omega,\mathcal{F},P)}$ and $\|f\|_{L^2(\mathbb{R}_+\times\Omega,\mathcal{B}(\mathbb{R}_+)\times\mathcal{F},dtdP)}$ are coincide. Using similar reasoning as in proving (8.14) one can derive for $f, g \in S_0^2$ that

$$\mathbf{E} I(f) I(g) = \mathbf{E}\int_0^\infty f_s g_s \, ds. \tag{8.15}$$

Equalities (8.13)-(8.15) show that the stochastic integral defined in (8.12) is a *linear* and *isometric* operator.

For function $f \in S_0^2$ we define a stochastic integral over interval $[0,t)$, $t > 0$, as follows

$$I_t(f) = I(I_{[0,t)}f) = \int_0^t f_s \, dW_s. \tag{8.16}$$

It turns out, $(I_t(f))_{t \geq 0}$ is a *martingale* with respect to $(\mathcal{F}_t)_{t \geq 0}$.

To prove it we fix $t > 0$, and for simplicity, assume that $t = t_k$.

Then we have
$I_{[0,t)}f_s = \sum_{i=0}^{n-1} f_{t_i} I_{[t_i,t_{i+1})}(t)$ and $I_t(f) = \sum_{i=0}^{n-1} f_{t_i}(W_{t_{i+1}} - W_{t_i})$, which are $\mathcal{F}_t$-measurable. Further, if $s \leq t$ and $s = t_u$ and $t = t_k$, then

$$\mathbf{E}(I_t(f) - I_s(f) | \mathcal{F}_s) = \sum_{i=u}^{k-1} \mathbf{E}\left(f_{t_i} \mathbf{E}(W_{t_{i+1}} - W_{t_i} | \mathcal{F}_{t_i}) | \mathcal{F}_s\right) = 0.$$

For function $f \in S_0$ one can rewrite

$$I_t(f) = \sum_{i=0}^{n-1} f_{t_i}(W_{t_{i+1} \wedge t} - W_{t_i \wedge t}),$$

and notice that the process $(I_t(f))_{t \geq 0}$ is *continuous*.

Now we take a monotonic sequence (s_j) such that $0 \leq s_0 \leq s_1 \leq \ldots \leq s_m < \infty$. Then the sequence $(I_{s_j}(f), \mathcal{F}_j)_{j=0,\ldots,m}$ is a martingale for which the Doob inequality is true

$$\mathbf{E} \sup_{0 \leq j \leq m} I_{s_j}^2(f) \leq 4\mathbf{E} I_{s_m}^2(f) \leq 4\mathbf{E}\int_0^{s_m} f_t^2 \, dt \leq 4\mathbf{E}\int_0^\infty f_t^2 \, dt. \tag{8.17}$$

Considering (s_j) rational and using the continuity of the trajectories of $I_t(f)$, we arrive to the *Doob inequality* for $f \in S_0^2$ taking *sup* over $t \in \mathbb{R}_+$ in (8.17):

8.2 Stochastic integrals with respect to a Wiener process

$$\mathbf{E} \sup_t I_t^2(f) \leq 4 \mathbf{E} \int_0^\infty f_t^2 dt. \tag{8.18}$$

One can extend the operator $I_t(f)$ from S_0^2 to its closure S^2. Any function $f \in S^2$ can be approximated by a sequence $f_n \in S_0^2$.

A concrete construction of such a sequence $(f_n)_{n=1,2,\ldots} \in S_0^2$ can be provided as follows. For any function $f \in S^2$ we define

$$f_n(t) = \begin{cases} n \int_{(k-1)/n}^{k/n} f(s)ds, & \text{for } t \in [k/n, (k+1)/n), \ k=1,2,\ldots,n^2-1, \ n=1,2,\ldots \\ 0, & \text{otherwise.} \end{cases}$$

Using the Jensen inequality we can find that

$$\int_{k/n}^{(k+1)/n} |f_n(t)|^2 dt = n \left| \int_{(k-1)/n}^{k/n} f(t)dt \right|^2 \leq \int_{(k-1)/n}^{k/n} |f(t)|^2 dt, \ k = 1, 2, \ldots$$

Hence, (a.s.) $\int_0^\infty |f_n(t)|^2 dt \leq \int_0^\infty |f(t)|^2 dt < \infty$,

$$\int_0^\infty |f(t) - f_n(t)|^2 dt \leq 2 \int_0^\infty \left(|f(t)|^2 + |f_n(t)|^2 \right) dt \leq 4 \int_0^\infty |f(t)|^2 dt < \infty.$$

Now, for arbitrary $N = 1, 2, \ldots$ we easily observe that $\mathbf{E} \int_0^n |f - f_n(t)|^2 dt \to 0$, $n \to \infty$, by construction of $(f_n)_{n=1,2,\ldots}$. Further, we have

$$\int_0^\infty |f(t) - f_n(t)|^2 dt = \int_0^N |f(t) - f_n(t)|^2 dt + \int_N^\infty |f(t) - f_n(t)|^2 dt$$

$$\leq \int_0^N |f(t) - f_n(t)|^2 dt + 2 \int_N^\infty \left(|f(t)|^2 + |f_n(t)|^2 \right) dt$$

$$\leq \int_0^N |f(t) - f_n(t)|^2 dt + 2 \int_N^\infty \left(|f(t)|^2 + \int_{N-1/n}^\infty |f_n(t)|^2 \right) dt$$

$$\leq \int_0^N |f(t) - f_n(t)|^2 dt + 2 \int_{N-1}^\infty |f(t)|^2 dt. \tag{8.19}$$

The term $\int_{N-1}^\infty |f(t)|^2 dt \to 0$ as $N \to \infty$, and by the dominated convergence theorem $\mathbf{E} \int_0^\infty |f(t) - f_n(t)|^2 dt \to 0$ as $n \to \infty$.

Then for m and $n \in \mathbb{Z}_+$ we have

$$\|I(f_m) - I(f_n)\|_{L^2(\Omega)} = \|I(f_m - f_n)\|_{L^2(\Omega)} = \|f_m - f_n\|_{S^2}.$$

It means that $(I(f_n))_{n \geq 1}$ is a fundamental sequence in a complete space $L^2(\Omega, \mathcal{F}, P)$. Hence, it converges to some element of this space:

$$I(f) = \lim_{n \to \infty} I(f_n), \tag{8.20}$$

and we call $I(f)$ a *stochastic integral* of $f \in S^2$. We also define for $t \geq 0$

$$\int_0^t f_s dW_s = \int_0^\infty I_{[0,t)}(s) f_s dW_s, \qquad (8.21)$$

and we prove that the stochastic integral (8.21) admits a *continuous modification*.

To prove it we take a sequence $f_n = f_n(t, \omega) \in S_0^2$ such that

$$\mathbf{E} \int_0^\infty |f(s) - f_n(s)|^2 ds \leq 2^{-n}, \; n = 1, 2, \ldots,$$

For each $t \geq 0$ we have in space $L^2(\mathbb{R}_+ \times \Omega, \mathcal{B}(0, \infty) \times \mathcal{F}, dt dP)$ that

$$I_{[0,t)}(s)(f) = I_{[0,t)}(s) f_1 + \ldots + I_{[0,t)}(s)(f_{n+1} - f_n) + \ldots, \qquad (8.22)$$

The equality (8.22) by continuity in the sense of this space is transformed to the next one

$$\int_0^t f(s) dW_s = \int_0^t f_1(s) dW_s + \ldots + \int_0^t |f_{n+1}(s) - f_n(s)| dW_s + \ldots, \qquad (8.23)$$

In the right hand side of (8.23) each term is continuous as functions $f_n \in S_0^2$. Therefore, to prove the continuity of stochastic integral (8.21) it is enough to provide a uniform convergence (a.s.) of series (8.21). First, due to the Doob inequality we obtain that

$$\mathbf{E} \sup_{t \geq 0} \left| \int_0^t (f_{n+1} - f_n(s)) dW_s \right|^2 \leq 4 \mathbf{E} \int_0^\infty (f_{n+1}(s) - f_n(s))^2 ds \qquad (8.24)$$

$$\leq 16 \cdot 2^{-n}. \qquad (8.25)$$

Second, combining (8.24) with the Chebyshev inequality, we get

$$P \left(\sup_t \left| \int_0^t (f_{n+1}(s) - f_n(s)) dW_s \right| \geq n^{-2} \right) \leq 16 n^4 2^{-n}. \qquad (8.26)$$

Application of (8.25) together with the Borel-Cantelli lemma certifies that the series (8.23) converges (a.s.).

Finally, we can conclude that the extension of $I_t(f)$ from space S_0^2 to S^2 leaves its *basic properties:*

(1) Linearity,
(2) Isometry,
(3) Martingality,
(4) Continuity of trajectories.

Before starting to state other properties of stochastic integrals we introduce a special class of random variables needed here.

8.2 Stochastic integrals with respect to a Wiener process

Let τ be a random variable taking values in $[0, \infty]$. We call it a *stopping time* on a stochastic basis $(\Omega, \mathcal{F}, (\mathcal{F})_{t \geq 0}, P)$, generated by a Wiener process $(W_t)_{t \geq 0}$, if $\{\omega : \tau(\omega) \leq t\} \in \mathcal{F}_t$ for every $t \in \mathbb{R}_+$.

A natural *example* is the *first hitting time* of the point a by W_t:

$$\tau = \tau_a = \inf\{t : W_t \geq a\},$$

$$\tau = \infty, \text{ if the set in brackets } \{.\} \text{ is empty.}$$

Indeed, one can observe that

$$\{\omega : \tau(\omega) \leq t\} = \Omega \setminus \{\omega : \tau(\omega) > t\}. \tag{8.27}$$

Further, we have a relation

$$\{\omega : \tau(\omega) > t\} = \{\omega : \max_{s \leq t} W_s(\omega) < a\}. \tag{8.28}$$

Let us note that

$$\max_{s \leq t} W_s = \sup_{r \leq t, r \in \mathbb{Q}} W_r, \tag{8.29}$$

where $\mathbb{Q}$ is the set of rational numbers, and hence $\max_{s \leq t} W_s$ is $\mathcal{F}_t$-measurable. Using continuity of trajectories of W_t and relations (8.27)-(8.29) we arrive to conclusion that $\{\omega : \tau(\omega) \leq t\} \in \mathcal{F}_t$.

Let us connect a stochastic integral $I_t(f) = X_t = \int_0^t f_s dW_s$, $f \in S^2$, and a stopping time τ. We have the two following equalities which are very useful:

$$X_\tau = \int_0^\infty I_{s<\tau} f_s dW_s = \int_0^\infty I_{s \leq \tau} f_s dW_s \ (a.s.), \tag{8.30}$$

$$\mathbf{E}\left(\int_0^\tau f_s dW_s\right)^2 = \mathbf{E}\int_0^\tau f_s^2 ds \text{ (Wald's identity)}, \tag{8.31}$$

To prove (8.30) we start with a stopping time τ taking discrete values $t_1, t_2, \ldots$ and find that $I_{s<t_k} f_s = I_{s<\tau} f_s$ on the set $\{\omega : \tau = t_k\}$.

On each such set (a.s.)

$$X_\tau = X_{t_k} = \int_0^\infty I_{(s<t_k)} f_s dW_s = \int_0^\infty I_{(s<\tau)} f_s dW_s.$$

Hence, it is true (a.s.) on $\Omega = \cup_k \{\tau = t_k\}$. To prove (8.30) for an arbitrary stopping time τ we approximate τ by a sequence of *discrete* stopping times

$$\tau_n = 2^{-2}[2^n \tau] + 2^{-n},$$

such that $\tau \leq \tau_n$ and $\tau_n - \tau \leq 2^{-n}$. Further, due to a continuity of X_t we get convergence $X_{\tau_n} \to X_\tau$ (a.s.) and by the dominated convergence theorem

$$\mathbf{E}\left(\int_0^\infty I_{\{s<\tau\}} f_s dW_s - \int_0^\infty I_{s<\tau_n} f_s dW_s\right)^2 = \mathbf{E}\int_0^\infty I_{\{\tau \le s<\tau_n\}} f_s^2 ds \to 0, \ n \to \infty.$$

Therefore, we obtain (8.30), and using isometry property we get (8.31):

$$\mathbf{E} X_\tau^2 = \mathbf{E}\int_0^\infty I_{s<\tau} f_s^2 ds = \mathbf{E}\int_0^\tau f_s^2 ds.$$

It turns out a certain extension of the notion of a stochastic integral with respect to a Wiener process can be done using the following *localization procedure*. It is given with the help of stopping times.

Consider the set of $\mathcal{B}(0,\infty) \times \mathcal{F}$−measurable adapted random functions $f = f_t(\omega)$ such that

$$\int_0^T f_s^2 ds < \infty \ \text{(a.s.) for all } T \in \mathbb{R}_+. \tag{8.32}$$

The family of such functions satisfying (8.32) is denoted by S.

We define a sequence of stopping time $(\tau_n)_{n=1,2,\ldots}$ such that

$$\tau_n = \inf\left(t : \int_0^t f_s^2 ds \ge n\right).$$

One can easily observe that (a.s.)

$$\int_0^{\tau_n} f_s^2 ds \le n \quad \text{and} \quad \tau_n \uparrow \infty, \ n \uparrow \infty, \quad \text{and} \quad I_{\{t<\tau_n\}} f \in S^2 \quad \text{for} \quad b = 1, 2, \ldots.$$

It follows from here that

$$X_t^n = \int_0^t I_{\{s<\tau_n\}} f_s dW_s,$$

is well defined for $n = 1, 2, \ldots$ one can also note that

$$X_t^n(\omega) = X_t^m(\omega),$$

for $m \ge n$, $0 \le t \le \tau_n(\omega)$ and $\omega \in \Omega'$, $P(\Omega') = 1$.

Moreover, the sequence $(X_t^n)_{n=1,2,\ldots}$ converges uniformly on finite time intervals.

All arguments above give a possibility to define a stochastic integral for each $f \in S$ as follows

$$\int_0^t f_s dW_s = \lim_{n\to\infty} \int_0^t I_{s<\tau_n} f_s dW_S. \tag{8.33}$$

Stochastic integral (8.33) for $f \in S$ admits a continuous modification and saves a linearity property. But it is not a martingale anymore. Nevertheless, the martingale property is fulfilled *locally*:

$$\left(\int_0^{t\wedge\tau_n} f_s dW_s\right)_{t\ge 0} \text{ is a square-integrable martingale for } n = 1, 2, \ldots$$

8.2 Stochastic integrals with respect to a Wiener process

Remark 8.3 Let us make a remark regarding *finite-dimensional processes*.

We say that (W_t) is a d–dimensional Wiener process if $W_t = (W_t^1, \ldots, W_t^d)$, where (W_t^i), $i = 1, \ldots, d$, are one-dimensional independent Wiener process.

For d–dimensional random function $f_t = (f_t^1, \ldots, f_t^d)$ with components from S we define

$$\int_0^t f_s dW_s = \sum_{i=1}^d \int_0^t f_s^i dW_s^i.$$

As far as a stochastic integration of matrix-valued random functions $\sigma_t = (\sigma_t^i)$, $i = 1, \ldots, d$, $j = 1, \ldots, d$, and $\sigma^{ij} \in S$ we define $\int_0^t \sigma_s dW_s$ as a d–dimensional process with the i–th coordinate $\sum_{j=1}^d \int_0^t \sigma_s^{ij} dW_s^j$.

Remark 8.4 The construction of a stochastic integral developed in this section was given by K. Ito. That is why this integral is often called the *Ito integral*. A similar construction was given by R. Stratonovich who proposed to use a different class of integral sums:

$$\sum_{i=0}^{N-1} f(t_i^*)(W_{t_{i+1}} i W_{t_i}),$$

where $t_i^* = \frac{1}{2}(t_{i+1} + t_i)$.

The corresponding stochastic integral is called the *Stratonovich integral*.

Remark 8.5 It is well-known that the construction of the Rieman integral as a limit of the integral sums does not depend on the choice of intermediate points inside of subdivision intervals. It was shown in Remark 8.4 that the construction of a stochastic integral is not invariant with respect to such a choice. Let us give another example of this type. Take $f(t) = W(t) = W_t$ and consider a subdivision of $[0, T]$:

$$0 = t_0 < t_1 < \ldots < t_n = T, \quad t_j = \frac{Tj}{n}, \quad j = 1, \ldots, n.$$

For this subdivision we define two integral sums as follows

$$I_1^{(n)} = \sum_{i=1}^n W_{t_{i-1}}(W_{t_i} - W_{t_{i-1}}) \text{ and } I_2^{(n)} \sum_{i=1}^n W_{t_i}(W_{t_i} - W_{t_{i-1}}).$$

For two real numbers a and b we note that $a(b-a) = \frac{1}{2}(b^2 - a^2) - \frac{1}{2}(a-b)^2$ and $b(b-a) = \frac{1}{2}(b^2 - a^2) + \frac{1}{2}(a-b)^2$.

Applying these elementary equalities to the integral sums we obtain that

$$I_1^{(n)} = \frac{1}{2}\sum_{i=1}^n (W_{t_i}^2 - W_{t_{i-1}}^2) - \frac{1}{2}\sum_{i=1}^n (W_{t_i} - W_{t_{i-1}})^2 = \frac{1}{2}W_T^2 - \frac{1}{2}\sum_{i=1}^n (W_{t_i} - W_{t_{i-1}})^2,$$

$$I_2^{(n)} = \frac{1}{2}\sum_{i=1}^n (W_{t_i}^2 - W_{t_{i-1}}^2) - \frac{1}{2}\sum_{i=1}^n (W_{t_i} - W_{t_{i-1}})^2 = \frac{1}{2}W_T^2 + \frac{1}{2}\sum_{i=1}^n (W_{t_i} - W_{t_{i-1}})^2.$$

Hence,

$$\mathbf{E}I_1^{(n)} = \frac{1}{2}T - \frac{1}{2}\mathbf{E}\sum_{i=1}^{n}(W_{t_i} - W_{t_{i-1}})^2 \to \frac{T}{2} - \frac{T}{2} = 0, \ n \to \infty,$$

$$\mathbf{E}I_2^{(n)} = \frac{1}{2}T + \frac{1}{2}\mathbf{E}\sum_{i=1}^{n}(W_{t_i} - W_{t_{i-1}})^2 \to \frac{T}{2} + \frac{T}{2} = T, \ n \to \infty,$$

and we observe that L^2-limits of integral sums are different.

Remark 8.6 It is very often another type of measurability of f is exploited in the definition of stochastic integrals, i.e. f is *progressively measurable* in the sense: for each t

$$f : ([0,t] \times \Omega, \mathcal{B}(0,t) \times \mathcal{F}_t) \to (\mathbb{R}, \mathcal{B}(\mathbb{R})).$$

Without loss of generality we can count S_0^2, S^2, S as classes of progressively measurable random functions.

8.3 The Ito processes: Formula of changing of variables, theorem of Girsanov, representation of martingales

Let us take two progressively measurable (or $\mathcal{F}_t$-adapted and $\mathcal{B}(0,\infty) \times \mathcal{F}$-measurable) $(b_t(\omega))_{t \geq 0}$ and $(\sigma_t(\omega))_{t \geq 0}$ satisfying the next integrability conditions: for each $T > 0$ (a.s.)

$$\int_0^T |b_s(\omega)|ds < \infty, \quad \int_0^T |\sigma_s(\omega)|^2 ds < \infty. \tag{8.34}$$

Conditions (8.34) imply a possibility to define a new stochastic process

$$X_t = X_0 + \int_0^t b_s ds + \int_0^t \sigma_s dW_s, \tag{8.35}$$

where X_0 is a finite (a.s.) random variable, $(W_t)_{t \geq 0}$ is a Wiener process generating filtration $(\mathcal{F}_t)_{t \geq 0}$.

The process $(X_t)_{t \geq 0}$ is called the *Ito process*. It is very convenient to write equality (8.35) in the following differential form

$$dX_t = b_t dt + \sigma_t dW_t. \tag{8.36}$$

In the above representation (8.36) we can speak about dX_t as a *stochastic differential* of $(X_t)_{t \geq 0}$.

We will study the class of the Ito processes using both form (8.35) and (8.36). As the first result playing a very important role in Stochastic Analysis is the *formula of changing of variables* (the Ito formula) for the class of smooth functions $F(t, x)$,

8.3 The Ito processes: Formula of changing of variables ...

having one continuous derivative $\frac{\partial F}{\partial t}$ and two continuous derivatives $\frac{\partial F}{\partial x}$ and $\frac{\partial^2 F}{\partial x^2}$, i.e. $F \in C^{1,2}$.

We start with the next lemma.

Lemma 8.3 *Let $(W_t)_{t \geq 0}$ be a Wiener process, then for each $t \geq 0$ (a.s.)*

$$W_t^2 = 2 \int_0^t W_s dW_s + t. \tag{8.37}$$

Proof Without loss of generality we consider only $t = 1$ and find that in the sense of convergence in probability (P–lim):

$$\int_0^t W_s dW_s = \lim_{n \to \infty} \sum_{k<n} W_{\frac{k}{n}} \left(W_{\frac{k+1}{n}} - W_{\frac{k}{n}} \right)$$

$$= \lim_{n \to \infty} \left[\sum_{k<n} \frac{1}{2} \left(W_{\frac{k+1}{n}} + W_{\frac{k}{n}} \right) \left(W_{\frac{k+1}{n}} - W_{\frac{k}{n}} \right) - \sum_{k<n} \frac{1}{2} \left(W_{\frac{k+1}{n}} - W_{\frac{k}{n}} \right)^2 \right]. \tag{8.38}$$

It is clear that as $n \to \infty$

$$\sum_{k<n} \frac{1}{2} \left(W_{\frac{k+1}{n}} + W_{\frac{k}{n}} \right) \left(W_{\frac{k+1}{n}} - W_{\frac{k}{n}} \right) = \frac{1}{2} \sum_{k<n} \left(W_{\frac{k+1}{n}}^2 - W_{\frac{k}{n}}^2 \right) \to \frac{1}{2} W_1^2,$$

$$\sum_{k<n} \left(W_{\frac{k+1}{n}} - W_{\frac{k}{n}} \right)^2 \to \frac{1}{2}.$$

Hence, we can get from (8.38) the equality (8.37). □

The next step is the formula for the product of two Ito's processes $(X_t^i)_{t \geq 0}$, $i = 1, 2$, with coefficients b^i, σ^i correspondingly.

Lemma 8.4 *For each $t \geq 0$ (a.s.)*

$$dX_t^1 X_t^2 = X_t^1 dX_t^2 + X_t^2 dX_t^1 + \sigma_t^1 \sigma_t^2 dt. \tag{8.39}$$

Proof Dividing $[0, t]$ with the help of a subdivision $(t_k)_{k \leq n}$ with the diameter $\max_{k \leq n} |t_k - t_{k-1}| \to 0$, $n \to \infty$, we can reduce the proof to the "small" intervals (t_k, t_{k+1}) of this subdivision.

For such small enough interval we have a discrete analog of (8.39):

$$X_{t_{k+1}}^1 X_{t_{k+1}}^2 - X_{t_k}^1 X_{t_k}^2 = \int_{t_k}^{t_{k+1}} X_s^1 \left[b_s^2 ds + \sigma_s^2 dW_s \right]$$

$$+ \int_{t_k}^{t_{k+1}} X_s^2 \left[b_s^1 ds + \sigma_s^1 dW_s \right] + \int_{t_k}^{t_{k+1}} \sigma_s^1 \sigma_s^2 ds,$$

where we can think about b^i, σ^i, $i = 1, 2$ as constants. Therefore, the problem is reduced to the following four cases:

(1) $X_t^1 = t$, $X_t^2 = t$,

(2) $X_t^1 = t$, $X_t^2 = W_t$,
(3) $X_t^1 = W_t$, $X_t^2 = t$,
(4) $X_t^1 = W_t$, $X_t^2 = W_t$.

Case (1) is the usual differentiation case.

Case (4) is Lemma 8.3.

Case (2)-(3) are equivalent, and we need to show that for each $t \geq 0$ (a.s.)

$$tW_t = \int_0^t W_s ds + \int_0^t s dW_s,$$

which is true due to the following obvious relation: for $t = 1$

$$1 \cdot W_1 = P - \lim_{n \to \infty} \sum_{k<n} \left[\frac{k+1}{n} W_{\frac{k+1}{n}} - \frac{k}{n} W_{\frac{k}{n}} \right]$$

$$= P - \lim_{n \to \infty} \left[\sum_{k<n} \frac{k}{n} \left(W_{\frac{k+1}{n}} - W_{\frac{k}{n}} \right) + \sum_{k<n} \frac{1}{n} W_{\frac{k+1}{n}} \right].$$

So, we arrive to the equality (8.39). $\square$

Lemma 8.5 *For any polynomial function $P_n(x)$ of degree $n = 1, 2, \ldots$ we have*

$$dP_n(W_t) = P_n'(W_t) dW_t + \frac{1}{2} P_n''(W_t) dt. \tag{8.40}$$

Proof It is sufficient to prove (8.40) for $P_n(x) = x^n$. We can use the induction method to prove that

$$dW_t^n = nW_t^{n-1} dW_t + \frac{n(n-1)}{2} W_t^{n-2} dt. \tag{8.41}$$

For $n = 1$ the formula (8.41) is trivial. Further, we take $X_t^1 = W_t^n$ and $X_t^2 = W_t$, use Lemma 8.4 with

$$b_t^1 = \frac{n(n-1)}{2} W_t^{n-1}, \quad \sigma_t^1 = nW_t^{n-2}, \quad b_t^2 = 0, \quad \sigma_t^2 = 1,$$

and obtain (8.41):

$$dW_t^{n+1} = W_t^n dW_t + W_t \left[nW_t^{n-1} dW_t + \frac{n(n-1)}{2} W_t^{n-2} dt \right]$$
$$+ nW_t^{n-1} dt$$
$$= (n+1) W_t^n dW_t + \frac{n(n+1)}{2} W_t^{n-1} dt.$$

$\square$

Corollary 8.1 *If $F = F(x)$ is a continuously differentiable function with continuous derivatives $F'(x)$ and $F''(x)$, then for $t \geq 0$ (a.s.)*

8.3 The Ito processes: Formula of changing of variables ...

$$dF(W_t) = F'(W_t)dW_t + \frac{1}{2}F''(W_t)dt. \qquad (8.42)$$

To prove (8.42) we uniformly (on compact intervals) approximate $F(x)$ by a sequence of polynomials P_n such that

$$F(x) = \lim_{n\to\infty} P_n(x), F'(x) = \lim_{n\to\infty} P'_n(x), F''(x) = \lim_{n\to\infty} P''_n(x)$$

and take the limit in the formula (8.40).

Corollary 8.2 *Assume $F : \mathbb{R}_+ \times \mathbb{R} \to \mathbb{R}$ is a function of class $C^{1,2}$, then for $t \geq 0$ (a.s.)*

$$dF(t, W_t) = \left[\frac{\partial F}{\partial t}(t, W_t) + \frac{1}{2}\frac{\partial^2 F}{\partial x^2}(t, W_t)\right]dt + \frac{\partial F}{\partial x}(t, W_t)dW_t. \qquad (8.43)$$

The first step of the proof is the case $F(t, x) = F_1(t)F_2(x)$. Then (8.43) follows from Lemma 8.4 and Corollary 8.1. The general case of F can be derived by approximation of $F(t, x)$ by a sequence of functions $F_n(t, x) = \sum_{k\leq n} F_{1,n}(t)F_{2,n}(x)$ with $\frac{\partial F_n}{\partial t}, \frac{\partial F_n}{\partial x}, \frac{\partial^2 F_n}{\partial x^2}$ converge uniformly on compact sets.

Remark 8.7 Let us define the local time of W at the level a during $[0, t]$:

$$l_W(t, a) = \lim_{\epsilon \to 0} \frac{1}{2\epsilon} \int_0^t I_{(a-\epsilon, a+\epsilon)}(W_s)ds,$$

which exists (a.s.) and in the sense of L^2-space. Using this notion, one can get a version of the Ito formula for non-differentiable functions, say, $F(x) = |x|$:

$$|W_t - a| - |W_0 - a| = \int_0^t sgn(W_s - a)dW_s + l_W(t, a),$$

where

$$sgn(x) = \begin{cases} 1, & x > 0, \\ 0, & x = 0, \\ -1, & x < 0. \end{cases}$$

Such a formula is called as the Tanaka formula.

The final formula (for Ito's processes) is given in the next theorem.

Theorem 8.2 *Assume $F \in C^{1,2}$ and*

$$dX_t = b_t(\omega)dt + \sigma_t(\omega)dW_t.$$

Then for $t \geq 0$ (a.s.)

$$dF(t, X_t) = \left[\frac{\partial F}{\partial t}(t, X_t) + b_t\frac{\partial F}{\partial x}(t, X_t) + \frac{\sigma_t^2}{2}\frac{\partial^2 F}{\partial x^2}(t, X_t)\right]dt + \sigma_t\frac{\partial F}{\partial x}(t, X_t)dW_t.$$
$$(8.44)$$

The formula (8.44) admits a generalization to a multidimensional case. Namely, assume $F : \mathbb{R}_+ \times \mathbb{R}^k \to \mathbb{R}$ of class $C^{1,2}$, the process (X_t) in (8.35) is defined by a vector-function $b_t(\omega) = b(t, \omega) = (b_t^1(\omega), \ldots, b_t^k(\omega))$ and a matrix-value function $\sigma_t(\omega) = \sigma(t, \omega) = (\sigma_t^{ij}(\omega))_{i \le k, j \le d}$:

$$X_t^i = X_0^i + \int_0^t b_s^i ds + \int_0^t \sum_{j=1}^d \sigma_s^{ij} dW_s^j, \quad i = 1, \ldots, k,$$

and $W_t = (W_t^1, \ldots, W_t^d)$ is a d–dimensional Wiener process. Then for $t \ge 0$ (a.s.)

$$dF(t, X_t) = \left[\frac{\partial F}{\partial t}(t, X_t) + \sum_{i=1}^k \frac{\partial F}{\partial x^i}(t, X_t) b_t^i + \frac{1}{2} \sum_{i,j=1}^k \frac{\partial^2 F}{\partial x_i \partial x_j}(t, X_t) \right. \quad (8.45)$$

$$\left. \times \sum_{l=1}^d \sigma_t^{il} \sigma_t^{lj} \right] dt + \sum_{j=1}^d \left[\sum_{i=1}^k \frac{\partial F}{\partial x_i}(t, X_t) \sigma_t^{ij} \right] dW_t^j. \quad (8.46)$$

Example 8.2 (Stochastic exponential) Consider the Ito process (8.35). Define the process

$$\mathcal{E}_t(X) = \exp\left\{ X_t - \frac{1}{2} \int_0^t \sigma_s^2 ds \right\}. \quad (8.47)$$

Applying the Ito formula to function $F(t, x) = e^{-\frac{\sigma^2}{2} t} e^x$, $(t, x) \in \mathbb{R}_+ \times \mathbb{R}$, we find that

$$d\mathcal{E}_t(X) = \mathcal{E}_t(X) dX_t, \quad \mathcal{E}_0(X) = 1. \quad (8.48)$$

We will call $(\mathcal{E}_t(X))_{t \ge 0}$ a *stochastic exponent* of process (X_t).

Let us count the following properties of stochastic exponentials (8.47)-(8.48).

Problem 8.3 1. $\mathcal{E}_t(X) > 0$ (a.s.), $t \in \mathbb{R}_+$;
2. $\frac{1}{\mathcal{E}_t(X)} = \mathcal{E}_t(\tilde{X})$, (a.s.) $t \in \mathbb{R}_+$, where $\tilde{X}_t = X_t - \int_0^t \sigma_s^2 ds$;
3. If $b_t = 0$ a.s., $t \in \mathbb{R}_+$, then $(\mathcal{E}_t(X))_{t \ge 0}$ is a (local) martingale;
4. $\mathcal{E}_t(X^1)\mathcal{E}_t(X^2) = \mathcal{E}_t\left(X^1 + X^2 + [X^1, X^2]\right)$ (a.s.), $t \in \mathbb{R}_+$, where $dX_t^i = b_t^i dt + \sigma^i dW_t$, $d[X^1, X^2]_t = \sigma_t^1 \sigma_t^2 dt$, $i = 1, 2$, and this property is called a *multiplication rule* of stochastic exponentials.

We already demonstrated how important and useful the Ito formula is. The next consideration has an incredible meaning for Stochastic Analysis and its applications. Let us consider a d–dimensional random function $b_t = (b_t^1, \ldots, b_t^d)$ belonging to the space S (or S^2) and define a new process

8.3 The Ito processes: Formula of changing of variables ...

$$Z_t = \exp\left\{\int_0^t b_s dW_s - \frac{1}{2}\int_0^t |b_s|^2 ds\right\}$$
$$= \exp\left\{\sum_{i=1}^d \int_0^t b_s^i dW_s^i - \frac{1}{2}\sum_{i=1}^d \int_0^t \left(b_s^i\right)^2 ds\right\}, \quad (8.49)$$
$$Z_0 = 1.$$

The process (8.49) satisfies the following properties
(1) $dZ_t = Z_t b_t dW_t$, $Z_0 = 1$;
(2) $(Z_t)_{t\geq 0}$ is a supermartingale, which is a martingale, if $(b_t)_{t\geq 0}$ is bounded or $\mathbf{E}Z_t = 1$, $t \in \mathbb{R}_+$;
(3) if $\mathbf{E}Z_T = 1$ for some $T > 0$, then $(Z_t)_{t\leq T}$ is a martingale, and for any sequence $b_n \in S(S^2)$ with the property $\int_0^T |b_n - b|^2 ds \to 0$, $n \to \infty$, (a.s.) the corresponding sequence $Z_T(b_n)$ converges to $Z_T(b)$ in the space L^1.

To prove (1) we just apply the Ito formula. For the proof of (2) we consider a sequence of stopping times

$$\tau_n = \inf\left(t : \int_0^t |b_s|^2 Z_s^2 ds \geq n\right)$$

and find that the process

$$\left(\int_0^t I_{(s<\tau_n)} b_s Z_s dW_s\right)_{t\geq 0}$$

is a martingale, as well as $(Z_{t\wedge\tau_n})_{t\geq 0}$.
For $t_1 \leq t_2$ we have (a.s.)

$$\mathbf{E}\left(Z_{t_2\wedge\tau_n}|\mathcal{F}_{t_1}\right) = Z_{t_1\wedge\tau_n}, \quad n=1,2,...$$

and as $n \to \infty$ we get by Fatou's Lemma that

$$\mathbf{E}(Z_{t_2}|\mathcal{F}_{t_1}) \leq Z_{t_1} \quad (a.s.),$$

which certifies that $(Z_t)_{t\geq 0}$ is a supermartingale, and therefore

$$\mathbf{E}\exp\left(\int_0^t b_s dW_s - \frac{1}{2}\int_0^t |b_s|^2 ds\right) \leq 1. \quad (8.50)$$

To prove (3) we first note that for $|b_s| \leq K < \infty$ we derive from (8.50) that

$$\mathbf{E}\int_0^t |b_s|^2 Z_s^2 ds \leq K^2 \mathbf{E} \int_0^t Z_s^2 ds$$
$$= K^2 \int_0^t \mathbf{E} Z_s^2(2b) \exp\left(\int_0^t |b_u|^2 du\right) ds$$
$$\leq K^2 \int_0^t e^{K^2 s} ds < \infty,$$

and conclude that $\left(\int_0^t b_s Z_s ds\right)_{t\geq 0}$ and $(Z_t)_{t\geq 0}$ are martingales.

The claim (3) follows from claim (2) above, where $\mathbf{E} Z_T(b) = 1$ and $Z_T(b^n) \xrightarrow{P} Z_T(b), n \to \infty$. As well as $\mathbf{E}|Z_T(b^n) - Z_T(b)| \to 0, n \to \infty$, and hence $Z_t(b^n) = \mathbf{E}(Z_T(b^n)|\mathcal{F}_t)$, $t \leq T$, a.s. Taking the limit in the above equality as $n \to \infty$ we get the martingale property of $(Z_t(b))_{t \leq T}$.

Now we are ready to formulate a *beautiful Girsanov Theorem*.

Theorem 8.3 *Assume $(W_t)_{t \leq T}$ is a Wiener process on a probability space $(\Omega, \mathcal{F}, P)$. Let b be a random function from space S (or S^2) such that $\mathbf{E} Z_T = 1$. Define process $X_t = W_t - \int_0^t b_s ds$. Then*

(1) a random variable Z_T defines a new probability measure $\tilde{P}$, equivalent to P, i.e. $\tilde{P} << P$ and $P << \tilde{P}$ with the density $Z_T = \frac{d\tilde{P}}{dP}$;

(2) The process (X_t) is a Wiener process with respect to $\tilde{P}$.

Proof. For simplicity, consider a bounded function (b_t). For $0 \leq t_0 \leq t_1 \leq \ldots \leq t_n = T$ and fixed $(\lambda_j)_{j=0,\ldots,n-1}$ from $\mathbb{R}^d$ we define the process λ_s as $i\lambda_j$ on $[t_j, t_{j+1})$. Further, we obtain that

$$\tilde{\mathbf{E}} \exp\left\{i \sum_{j=0}^{n-1} \lambda_j (X_{t_{j+1}} - X_{t_j})\right\} = \mathbf{E} \exp\left\{\int_0^T \lambda_s dW_s - \int_0^T \lambda_s b_s ds\right\} Z_T(b)$$
$$= \mathbf{E} Z_T(\lambda + b) \exp\left(\frac{1}{2}\int_0^T \lambda_s^2 ds\right)$$
$$= e^{1/2 \int_0^T \lambda_s^2 ds}$$

which corresponds to a Gaussian distribution. Taking into account $X_0 = 0$, we arrive to the claim (2) of the theorem.

To finish this section we study the structure of any martingale $(M_t)_{t \geq 0}$ on stochastic basis $(\Omega, \mathcal{F}, (\mathcal{F}_t = \mathcal{F}_t^W)_{t \geq 0}, P)$. It turns out, the set of all martingales on the stochastic basis, generated by a Wiener process $(W_t)_{t \geq 0}$ is determined by stochastic integrals. Namely, any square integrable martingale $(M_t)_{t \geq 0}$, $M_0 = 0$, admits

$$M_t = \int_0^t \phi_s dW_s \quad (a.s.), t \geq 0, \tag{8.51}$$

where $(\phi_t)_{t \geq 0}$ is uniquely defined progressively measurable function from space S^2. The relation (8.51) is called a martingale representation. Let us prove it for a finite

8.3 The Ito processes: Formula of changing of variables ...

interval $[0,T]$. For a deterministic function $\beta \in L^2(0,T)$ we define the Girsanov exponent

$$Z_t = Z_t(\beta) = exp\left\{\int_0^t \beta_u du - \frac{1}{2}\int_0^t \beta_u^2 du\right\}, Z_0 = 1$$

Let us note that

$$Z_t^2(\beta) = Z_t(2\beta)exp\left\{\int_0^t \beta_u^2 du\right\},$$

and $\mathbf{E}Z_t(2\beta) \leq 1$ for all $t \leq T$.

Hence, we obtain

$$\mathbf{E}Z_t^2(\beta) \leq exp\left\{\int_0^t \beta_u^2 du\right\} < \infty.$$

Let $\mathcal{X}^2$ be a class of random variables X from the space $L^2 = L^2(\Omega, \mathcal{F} = \mathcal{F}_T^W, P) = L^2$ such that $X = \int_0^T \phi_s dW_s$ with a progressively measurable ϕ and $\int_0^T \mathbf{E}\phi_s^2 ds < \infty$. Further, if Y is an arbitrary random variable from L^2 such that $\mathbf{E}Y = 0$ and $Y \perp \mathcal{X}^2$ in the L^2-sense, then $\mathbf{E}Y \cdot Z_T(\beta) = 0$ for all $\beta \in L^2(0,T)$. Therefore taking an arbitrary subdivision $(t_i), i = 1,...,n$ of the interval $[0,T]$ and constructing an arbitrary step function with (t_i), we obtain that $\mathbf{E}(Y|W_{t_1},...,W_{t_n}) = 0$. Hence, $Y = 0$ (a.s.) due to $\mathcal{F}_T = \mathcal{F}_T^W$. The last step of the proof is clear. We take M_T as $X \in \mathcal{X}^2$ and using the martingale and stochastic integral properties, obtain

$$M_t = \mathbf{E}(M_T|\mathcal{F}_t) = \int_0^t \phi_s dW_s.$$

Chapter 9
Stochastic differential equations, diffusion processes and their applications

Abstract The chapter presents stochastic differential equations (SDEs) and their connections with diffusion processes and partial differential equations (PDEs). The existence and uniqueness of solutions of SDEs are proved under Lipschitz's conditions. Two important processes (Geometric Brownian Motion (GBM) and the Ornstein-Uhlenbeck process) are constructed on this theoretical base. The difference between ordinary differential equations and SDEs are discussed. As a part of this discussion, the existence of a solution (weak solution) of any SDE with measurable bounded drift coefficient and unit diffusion is proved with the help of the miracle Girsanov theorem. Moreover, it is shown by mean of the method of monotonic approximations that such a solution will be strong if the grift coefficient is a bounded piece-wise smooth function. Diffusion processes are defined as Markov processes for which their transition densities satisfy the asymptotic properties of Kolmogorov. The backward and forward equations of Kolmogorov are derived. A connection between SDEs and PDEs are stated with the help of the Feynman-Kac theorem. Absolute continuity of distributions of diffusion processes is studied with the help of the Girsanov theorem. A special attention is paid to the class of controlled diffusion processes for which the Hamilton-Jacobi-Bellman optimality equation is derived. It is shown how the theory of diffusion processes and SDEs are helpful in mathematical finance (Bachelier and Black-Scholes models) and in statistics of random processes (see [3], [5], [14], [17], [21], [22], [23], [24], [25], [30], [35], [39], [41], [42], and [44]).

9.1 Stochastic differential equations

Let $b(t, x)$ and $\sigma(t, x)$ be (t, x)–measurable locally bounded functions from $\mathbb{R}_+ \times \mathbb{R} \to \mathbb{R}$, i.e. these functions are bounded on each compact set of $\mathbb{R}_+ \times R$.

If a continuous $\mathcal{F}_t$-adapted process $X(t, \omega) = X_t(\omega)$ such that for all $t \geq 0$ (a.s.)

9 Stochastic differential equations, diffusion processes and their applications

$$X_t(\omega) = X_0(\omega) + \int_0^t b(s, X_s)ds + \int_0^t \sigma(s, X_s)dW_s, \tag{9.1}$$

where $X_0(\omega)$ is a finite random variable, measurable w.r. to $\mathcal{F}_0$, $(W_t)_{t \geq 0}$ is a Wiener process, then it is called a *solution* of (9.1). We rewrite (9.1) in an equivalent differential form

$$dX_t = b_t(X_t)dt + \sigma_t(X_t)dW_t. \tag{9.2}$$

We say that such a solution is *unique* if for any other solution $(\tilde{X}_t)$ we have

$$P(\omega : \sup_t |X_t(\omega) - \tilde{X}_t(\omega)| > 0) = 0. \tag{9.3}$$

Theorem 9.1 *Assume that coefficients $b(t, x)$ and $\sigma(t, x)$ of stochastic equations* (9.1) (9.2) *satisfy the conditions:*
1) For each $T \geq 0$ there exists a constant k_T such that

$$|b(t, x)| + |\sigma(t, x)| \leq k_T(1 + |x|)$$

for $(t, x) \in [0, T] \times \mathbb{R}$;
2) For all $c > 0$ there exists a constant $l_c > 0$ such that

$$|b(t, x) - b(t, y)| + |\sigma(t, x) - \sigma(t, y)| \leq l_c |x - y|$$

for all $(t, x) \in [0, c] \times [-c, c]$.
Then the equation (9.1)-(9.2) *admits one and only one solution.*

Proof First, we note that conditions 1)-2) of this theorem are called a *linear growth* and a *local Lipschitz* conditions, correspondingly. Second, we give the proof for a standard Lipschitz condition ($l_c = l$) and for a finite time interval $[0, T]$, and for $X_0(\omega) = x \in \mathbb{R}$.

Let us prove a uniqueness of solutions. Assume $X(t, \omega)$ and $\tilde{X}(t, \omega)$ are two solutions with the initial condition $X(0, \omega) = \tilde{X}(0, \omega) = x$. Then we have

$$\tilde{X}(t) - X(t) = \int_0^t \left[b(s, \tilde{X}(s)) - b(s, X(s)) \right] ds + \int_0^t \left[\sigma(s, \tilde{X}(s)) - \sigma(s, X(s)) \right] dW_s.$$

Further, using the Cauchy-Schwartz inequality and the isometry property, we obtain for each $t \in [0, T]$ that

9.1 Stochastic differential equations

$$\mathbf{E}|\tilde{X}(t) - X(t)|^2 \leq 2\mathbf{E}\left(\int_0^t [b(s, \tilde{X}(s)) - b(s, X(s))]\, ds\right)^2 \quad (9.4)$$

$$+ 2\mathbf{E}\left(\int_0^t [\sigma(s, \tilde{X}(s)) - \sigma(s, X(s))]\, dW_s\right)^2$$

$$\leq 2t\mathbf{E}\int_0^t [b(s, \tilde{X}(s)) - b(s, X(s))]^2\, ds + 2\mathbf{E}\int_0^t [\sigma(s, \tilde{X}(s)) - \sigma(s, X(s))]^2\, ds$$

$$\leq 2(t+1)l^2 \int_0^t \mathbf{E}|\tilde{X}(s) - X(s)|^2\, ds. \quad (9.5)$$

To continue we need the following *Gronwall lemma:*

If $\phi(t) \geq 0$ is an integrable function satisfying inequality for all $t \geq 0$ and fixed constants A and B:

$$\phi(t) \leq A + B\int_0^t \phi(s)\, ds. \quad (9.6)$$

Then $\phi(t) \leq Ae^{Bt}$ for all $t \geq 0$.

To prove Grownwall lemma we just differentiate the logarithm of the right hand side of (9.6):

$$\left(\ln\left(A + B\int_0^t \phi(s)\, ds\right)\right)' = \frac{B\phi(t)}{A + B\int_0^t \phi(s)\, ds} \leq B,$$

$$\ln\left(A + B\int_0^t \phi(s)\, ds\right) \leq \ln A + Bt,$$

$$A + B\int_0^t \phi(s)\, ds \leq Ae^{Bt}.$$

In relations (9.5) the function $\phi(t) = \mathbf{E}|\tilde{X}(t) - X(t)|^2$ and $A = 0, B = 2(T+1)l^2$. Hence, by the Gronwall lemma we obtain $\mathbf{E}|\tilde{X}(t) - X(t)|^2 = 0$. We derive from here that for the set $\mathbb{Q}_T$ of rational numbers of $[0, T]$ that

$$P\left(\sup_{t \in \mathbb{Q}_T} |\tilde{X}(t) - X(t)| = 0\right) = 1.$$

Due to a continuity of processes $X(t)$ and $\tilde{X}(t)$ we have

$$1 = P(\sup_{\mathbb{Q}_T} |\tilde{X}(t) - X(t)| = 0) = P(\sup_{t \in [0,T]} |\tilde{X}(t) - X(t)| = 0),$$

which means that X and $\tilde{X}$ are equivalent.

To prove the *existence* of a solution of (9.1)-(9.2), we apply the Picard method of successive approximations:

$$X^0(t) = x,$$
$$X^n(t) = x + \int_0^t b(s, X^{n-1}(s))ds + \int_0^t \sigma(s, X^{n-1}(s))dW_s. \quad (9.7)$$

For the sequence of approximations (9.7) we get for $t \in [0, T]$ that

$$|X^n(t)|^2 \leq 3|x|^2 + 3\left|\int_0^t b(s, X^{n-1}(s))ds\right|^2 + 3\left(\int_0^t b(s, X^{n-1}(s))dW_s\right)^2. \quad (9.8)$$

Inequality (9.8) implies for $t \leq T$ that

$$\mathbf{E}|X^n(t)|^2 \leq 3|x|^2 + L_T \int_0^t \mathbf{E}|X^{n-1}(s)|^2 ds, \quad (9.9)$$

where L_T depends on k_T.

Taking $\sup_{k \leq n}$ in (9.9) we derive from Gronwall's Lemma that

$$\sup_{k \leq n} \mathbf{E}|X^k(t)|^2 \leq 3|x|^2 e^{TL_T}, \quad t \in [0, T]. \quad (9.10)$$

Further, we have that

$$X^{n+1}(t) - X^n(t) = \int_0^t \left[b(s, X^n(s)) - b(s, X^{n-1}(s))\right] ds$$
$$+ \int_0^t \left[\sigma(s, X^n(s)) - \sigma(s, X^{n-1}(s))\right] dW_s.$$

and applying a similar estimation as the proof of uniqueness, we get for $t \leq T$:

$$\mathbf{E}|X^{n+1}(t) - X^n(t)|^2 \leq K_T \int_0^t \mathbf{E}|X^n(s) - X^{n-1}(s)|^2 ds, \quad (9.11)$$

$$\mathbf{E}|X^1(t) - X^0(t)|^2 \leq \mathbf{E}\left(\int_0^t b(s, x))ds + \int_0^t \sigma(s, x)dW_s\right)^2 \leq C_T t, \quad (9.12)$$

with some constants C_T and K_T.

Using the induction we derive from (9.11)-(9.12) that

$$\mathbf{E}|X^{n+1}(t) - X^n(t)|^2 \leq C_T K_T^n \frac{t^{n+1}}{(n+1)!}. \quad (9.13)$$

Due to (9.13) we conclude that the series

$$X^0(t) + \sum_{n=0}^{\infty} \left(X^{n+1}(t) - X^n(t)\right)$$

converges uniformly to continuous process $X(t)$ in the L^2-sense.

To prove that this process is a solution of (9.1)-(9.2) we note for $t \in [0, T]$ that

9.1 Stochastic differential equations

$$\mathbf{E}\left(\int_0^t [b(s, X(s)) - b(s, X^n(s))]\, ds + \int_0^t [\sigma(s, X(s)) - \sigma X^n(s))]\, dW_s\right)^2$$
$$\leq K_T \int_0^T \mathbf{E}|X(s) - X^n(s)|^2 ds.$$

Using this inequality, we obtain that

$$\mathbf{E}\left(X(t) - x - \int_0^t b(s, X(s))ds + \int_0^t \sigma(s, X(s))dW_s\right)^2$$
$$\leq 2\mathbf{E}|X(t) - X^n(t)|^2$$
$$+ 2\mathbf{E}\left(\int_0^t b(s, X^{n-1}(s))ds + \int_0^t \sigma(s, X^{n-1}(s))dW_s - \int_0^t b(s, X(s))ds - \int_0^t \sigma(s, X(s))dW_s\right)^2$$
$$\leq 2\mathbf{E}|X(t) - X^n(t)|^2 + 2K_T \mathbf{E}|X(t) - X^{n-1}(t)|^2 \to 0,$$

as $n \to \infty$. □

Remark 9.1 To emphasize a dependence of the limit of successive approximations from an initial point x (correspondently, y) denote it by $X_x(t, \omega)$ (correspondently, $X_y(t, \omega)$). Similar to inequality (9.10) one can derive for $t \in [0, T]$:

$$\mathbf{E}|X_x(t) - X_y(t)|^2 \leq const(T)|x - y|^2 e^{t \cdot const(T)},$$
$$\mathbf{E}|X_x(t) - X_x(t + \Delta t)|^2 \leq const(T)|1 + |x|^2|^2 e^{t \cdot const(T)}.$$

Due to these inequalities the function $X_x(t)$ can be chosen measurable w.r. to x.

Example 9.1 Consider a linear SDE

$$dX_t = \mu X_t dt + \sigma X_t dW_t, \tag{9.14}$$

where $X_0 - \mathcal{F}_0$-measurable finite random variable. According to Theorem 8.1 the equation (9.14) has a unique solution. To find a concrete formula for X_t we rewrite (9.1) in the form

$$dX_t = (c + \sigma^2/2)X_t dt + \sigma X_t dW_t, \tag{9.15}$$

where $c = \mu - \sigma^2/2$.

Applying the Ito formula to $X_0 \exp\{ct + \sigma W_t\} = \tilde{X}_t$ we find that

$$d\tilde{X}_t = d\left(X_0 e^{ct + \sigma W_t}\right)$$
$$= \left(cX_0 e^{ct + \sigma W_t} + \frac{\sigma^2}{2} X_0 e^{ct + \sigma W_t}\right) dt + \sigma X_0 e^{ct + \sigma W_t} dW_t$$
$$= \left(c + \frac{\sigma^2}{2}\right) \tilde{X}_t dt + \sigma \tilde{X}_t dW_t.$$

So, using the uniqueness of solution of (9.14) we arrive to conclusion that $X_t = \tilde{X}_t$ (a.s.) for all $t \geq 0$. Hence,

$$X_t = X_0 e^{ct+\sigma W_t} = X_0 e^{(\mu-\sigma^2/2)t+\sigma W_t}, \qquad (9.16)$$

which is called a *Geometric Brownian Motion* (GBM). The GBM (9.16) is exploited as a stock price model in mathematical finance.

Example 9.2 Consider another linear stochastic differential equation

$$dX_t = -\alpha X_t dt + \sigma dW_t, \qquad (9.17)$$

X_0 is $\mathcal{F}_0$-measurable finite random variable, $\alpha \in \mathbb{R}$.

Let us transform (X_t) with the help of function $f(t,x) = e^{\alpha t}.x$, $x \in \mathbb{R}$. By the Ito formula we can find a stochastic differential of $Y_t = f(t, X_t)$:

$$dY_t = (f'_t - \alpha X_t f'_x + \frac{1}{2}\sigma^2 f''_{xx})dt + \sigma f'_x dW_t$$
$$= (\alpha e^{\alpha t} X_t - \alpha e^{\alpha t} X_t) dt + \sigma e^{\alpha t} dW_t = \sigma e^{\alpha t} dW_t.$$

It leads to $Y_t = X_0 + \sigma \int_0^t e^{\alpha s} dW_s$ and hence

$$X_t = e^{-\alpha t} Y_t = e^{-\alpha t} X_0 + \sigma e^{-\alpha t} \int_0^t e^{\alpha s} dW_s.$$

This is a famous *Ornstein-Uhlenbeck process*, which is called a *Vasicek model* for interest rate modeling in mathematical finance.

We observed here that for SDEs with Lipschitz-type coefficients existence and uniqueness of solutions are stated in a similar way as for ordinary differential equations. A natural question arises:

Are there new results in this theory, which are different from the deterministic theory?

We start our answer with the following *nice result of Girsanov*.

Theorem 9.2 *Let $b(t,x)$, $(t,x) \in [0,T] \times \mathbb{R}^d$, $d \geq 1$, be a bounded measurable function. Then there exists a probability space $(\Omega, \mathcal{F}, P)$, a continuous process (X_t) and a Wiener process on this space such that $(X_t)_{t \leq T}$ is a solution of the SDE*

$$dX_t = b(t, X_t)dt + dW_t, \ X_0 = 0, t \leq T. \qquad (9.18)$$

Proof Take any probability space $(\Omega, \mathcal{F}, \tilde{P})$ with a Wiener process $(X_t)_{t \leq T}$. Define another process

$$W_t = X_t - \int_0^t b(s, X_s)ds, \qquad (9.19)$$

and a new probability measure P with the density w.r.to $\tilde{P}$:

9.1 Stochastic differential equations

$$\frac{dP}{d\tilde{P}} = \exp\left(\int_0^T b(s, X_s)dX_s - \frac{1}{2}\int_0^T |b(s, X_s)|^2 ds.\right)$$

By the Girsanov theorem, the process $(W_t)_{t \le T}$ is a Wiener process on the probability space $(\Omega, \mathcal{F}, P)$. Hence, if we rewrite the equality (9.19) as

$$X_t = \int_0^t b(s, X_s)ds + W_t, \ t \le T,$$

we arrive to conclusion that $(X_t)_{t \le T}$ is a solution of (9.18). □

Remark 9.2 The result of Theorem 9.2 guarantees existence of a solution of (9.18) for any bounded measurable function $b(t, x)$. It is well-known that the ordinary differential equation $dx_t = b(t, x_t)dt$ does not admit a solution for any bounded measurable function $b(t, x)$.

Remark 9.3 Comparing solutions in Theorem 9.1 and 9.2 we can observe the following difference. The solution in Theorem 9.1 was constructed on *given probability space*. On the other hand, the solution in Theorem 9.2 was built on *some probability space*. They are naturally called *strong* and *weak* solutions, respectively, because any strong solution will be also a weak solution, but not vice versa.

Example 9.3 Let (β_t) be a Brownian Motion (BM) with $\beta_0 = y \in \mathbb{R}$. Define a process $B_t = \int_0^t sgn(\beta_s)d\beta_s$, where

$$sgn(x) = \begin{cases} 1, & x > 0, \\ -1, & x \le 0. \end{cases}$$

The process (B_t) is a BM by the Levy characterization. Further, we note that by a property of stochastic integral

$$y + \int_0^t sgn(\beta_s)dB_s = y + \int_0^t (sgn(\beta_s))^2 d\beta_s$$
$$= y + \int_0^t d\beta_s = y + \beta_t - y = \beta_t.$$

Hence, (β_t) solves the SDE

$$dX_t = sgn(X_t)dB_t, X_0 = y, \quad (9.20)$$

in the "weak" sense. Moreover, any other solution will be again a BM, and in the sense of its distribution such a solution is unique (weak uniqueness). But taking $y = 0$ we find that (β_t) and $(-\beta_t)$ solve the equation (9.20) with some BM (B_t) and $X_0 = 0$. It means there is no uniqueness as (9.5).

Let us develop the *method of monotonic approximations*, which is powerful enough to prove existence of strong solutions without the Lipschitz conditions. We start with the simplest version of the *comparison theorem* of strong solutions of SDEs.

Lemma 9.1 Assume that $b^i = b^i(x), x \in \mathbb{R}, i = 1, 2$, are bounded continuous functions, and $X^i, i = 1, 2$, are strong (continuous) solutions of the equations

$$dX_t^i = b^i(X_t^i)dt + dW_t, \quad X_0^i = 0, t \geq 0.$$

Then the inequality $b^1(x) < b^2(x)$ for all $x \in \mathbb{R}$ implies that $X_t^1(\omega) \leq X_t^2(\omega)$ for almost all $\omega \in \Omega$ and all $t \geq 0$.

Proof For each $\omega \in \tilde{\Omega}, P(\tilde{\Omega}) = 1$, we define

$$\theta = \begin{cases} \inf\{t > 0 : X_t^1(\omega) = X_t^2(\omega)\}, \\ \infty \text{ otherwise.} \end{cases}$$

Note that $X_t^2(\omega) - X_t^1(\omega) = \int_0^t \left[b^2(X_s^2(\omega)) - b^1(X_s^1(\omega)) \right] ds = \Delta_t(\omega)$. Due to continuity of $b^i(x), X_t^i(\omega), i = 1, 2$, and assumption $b^1(x) < b^2(x), x \in \mathbb{R}$, we find that $\Delta_0(\omega) > 0$, $\Delta_t(\omega)$ is continuous and $\Delta_t(\omega) > 0$ for all $t < \theta(\omega)$. If $\theta(\omega) = \infty$, then $X_t^2(\omega) = X_t^1(\omega)$, and Lemma is proved. Otherwise, we have $\theta(\omega) < \infty$ and $\Delta_\theta(\omega) = X_\theta^2 - X_\theta^1 = 0$. It gives a possibility to reproduce for θ the same consideration as for $t = 0$ and find that $X_t^2(\omega) - X_t^1(\omega) = \Delta_t(\omega) > 0$ for a neighborhood of $\theta(\omega)$, etc. Therefore, $X_t^2(\omega) - X_t^1(\omega) = \Delta_t(\omega) > 0$ for all $t \geq 0$. $\square$

Theorem 9.3 *Consider a homogeneous SDE (9.18) and suppose that the drift coefficient $b_t(x) = b(x)$ satisfies the following conditions:*
 (1) $b(x)$ has a finite number of points of discontinuity $\{x_1, \ldots, x_N\}$;
 (2) $b(x)$ is piece-wise smooth, i.e. for any $x \in \mathbb{R} \setminus \{x_1, \ldots, x_N\}$ the derivative $b'(x)$ exists;
 (3) $|b(x)|$ and $|b'(x)| \leq K < \infty$.
 Then the equation (9.18) has a strong solution.

Proof The first step is to show that $b(x)$ can be approximated by a sequence of functions $b_n(x)$ such that a strong solution of the equations

$$dX_n(t, \omega) = b_n(X_n(t, \omega))dt + dW(t, \omega), \qquad (9.21)$$

$$X_n(0, \omega) = 0, t \in [0, T], \quad n = 1, 2, \ldots,$$

does exist and is unique. Then the properties of these solutions are studied with the aim to show that the sequence $(X_n(t, \omega))_{n=1,2,\ldots}$ converges to a continuous $\mathcal{F}_t$-adapted process $X_\infty(t, \omega)$ which is the desired solution of the limit equation (9.18).

To construct the auxiliary sequence of functions $(b_n(x))_{n=1,2,\ldots}$ we take neighborhoods $V_r(x_i)$ of the points $x_1 < \ldots < x_N$ of radius $r \leq \frac{1}{4} \min_{1 \leq i \leq N-1} |x_{i+1} - x_i|$ and set

$$\tilde{b}_n(x) = b(x),$$
$$x \in \mathbb{R} \setminus \cup_{i=1}^N V_r(x_i).$$

9.1 Stochastic differential equations

It suffices now to build $\tilde{b}_n(x)$ in the neighborhood $V_r(x_i)$, $i = 1, 2, \ldots, N$. Consider $V_r(x_i)$ and let $b(x_i-) > b(x_i+)$ for limits from the left and right side of x_i correspondingly.

It follows from the smoothness of $b(x)$ that there exists a deterministic sequence $(\alpha_n^i)_{n=1,2,\ldots}$ such that $\alpha_n^i \in V_r(x_i)$, $\alpha_n^i > x_i$, $\alpha_n^i \downarrow x_i$ as $n \to \infty$, while the continuous function $\tilde{b}_n(x)$ is determined as follows

$$\tilde{b}_n(x) = b(x) \text{ for } x \in V_r(x_i) \cap \{(x_{i-1}, x_i) \cup [\alpha_n^i, x_{i+1})\},$$
$$\tilde{b}_n(x_i) = b(x_i-),$$

$\tilde{b}_n(x)$ is linear on $[x_i, \alpha_n^i]$ satisfying the condition $\tilde{b}_n(x) \geq b(x)$, $x \in (x_i, \alpha_n^i)$.

If $b(x_i-) < b(x_i+)$, the construction of $\tilde{b}_n(x)$ in $V_r(x_i)$ is analogous. Finally, we define $b_n(x) = \tilde{b}_n(x) + 1/n$, $n = 1, 2, \ldots$, with desirable properties for us:

$$b_n(x) > b_{n+1}(x), \ x \in \mathbb{R},$$
$$b_n(x) \downarrow b(x) \text{ as } n \to \infty, \ x \in \mathbb{R} \setminus \{x_1, \ldots, x_N\},$$

$b_n(x)$ satisfies a Lipschitz condition; and simultaneously for $n = 1, 2, \ldots$

$$|b_n(x)| \leq K + 1 < \infty.$$

Applying Theorem 9.1 to (9.21) with $b_n(x)$, constructed above, we get the existence (and uniqueness) of strong solutions $X_n(t)$, which are continuous and $\mathcal{F}_t$-adapted.

Further, according to Lemma 9.1 we obtain the system of inequalities

$$X_1(t, \omega) \geq X_2(t, \omega) \geq \ldots \geq X_n(t, \omega) \geq \ldots \ (a.s.), t \leq T.$$

Therefore, there exists a $\mathcal{F}_t$-adapted process $X_\infty(t, \omega)$ such that $X_\infty(t, \omega) = \lim_{n \to \infty} X_n(t, \omega)$ (a.s.), $t \in [0, T]$. □

Lemma 9.2 *The family of continuous functions* $(X_n(\cdot, \omega))_{n=1,2,\ldots}$, $\omega \in \tilde{\Omega}$, $P(\tilde{\Omega}) = 1$, *is uniformly bounded and equicontinuous.*

Proof For each ω from $\tilde{\Omega}$, $P(\tilde{\Omega}) = 1$, we have for all $n = 1, 2, \ldots$ that

$$|X_n(t, \omega)| \leq \left|\int_0^t b_n(X_n(s, \omega))ds\right| + |W(t, \omega)| \tag{9.22}$$

$$\leq \int_0^t |b_n(X_n(s, \omega))|ds + |W(t, \omega)| \tag{9.23}$$

$$\leq (K + 1)T + \max_{0 \leq t \leq T} |W(t, \omega)|, \tag{9.24}$$

which means that $(X_n(\cdot, \omega))_{n=1,2,\ldots}$ is uniformly bounded.

Further, for $\omega \in \tilde{\Omega}$, $P(\tilde{\Omega}) = 1, \delta > 0$, we obtain

$$\omega_{X_n(\cdot,\omega)}(\delta) = \sup_{s,t,|t-s|<\delta} |X_n(t,\omega) - X_n(s,\omega)|$$

$$\leq \sup_{s,t,|t-s|<\delta} \left| \int_s^t b_n(X_n(u,\omega))du \right| + |W_n(t,\omega) - W(s,\omega)|$$

$$\leq (K+1)\delta + \omega_{W(\cdot,\omega)}(\delta).$$

Hence, for almost all ω we have as $\delta \to 0$ that $\sup_n \omega_{X_n(\cdot,\omega)}(\delta) \leq (K+1)\delta + \omega_{W(\cdot,\omega)}(\delta) \to 0$.

In view of Lemma 9.2 and the Ascoli-Arzela test there exists (for $\omega \in \tilde{\Omega}, P(\tilde{\Omega}) = 1$) a subsequence $n'(\omega)$ such that

$$\sup_{0 \leq t \leq T} |X_{n'(\omega)}(t,\omega) - X_\infty(t,\omega)| \to 0, \ n'(\omega) \to \infty,$$

and therefore almost all paths of $X_\infty(t,\omega)$ are continuous on $[0,T]$.

Now we put $\phi_n(t,\omega) = b_n(W(t,\omega)), n = 1,2,\ldots, \phi_\infty(t,\omega) = b(W(t,\omega))$ and define probability measures $(P_n)_{n=1,2,\ldots}$ and P_∞ with the help of the Girsanov exponents:

$$\frac{dP_n}{dP} = \exp\left\{ \int_0^T \phi_n(s,\omega)dW(s,\omega) - \frac{1}{2}\int_0^T \phi_n^2(s,\omega)ds \right\},$$

$$\frac{dP_\infty}{dP} = \exp\left\{ \int_0^T \phi_\infty(s,\omega)dW(s,\omega) - \frac{1}{2}\int_0^T \phi_\infty^2(s,\omega)ds \right\}.$$

Problem 9.1 Prove that for $y \in \mathbb{R}$ and $t \in [0,T]$

$$P_n(\omega : W(t,\omega) \leq y) = P(\omega : X_n(t,\omega) \leq y), \quad (9.25)$$

$$P_n(\omega : W(t,\omega) \leq y) \xrightarrow[n\to\infty]{} P(\omega : W(t,\omega) \leq y), \quad (9.26)$$

$$P_n(\omega : X_n(t,\omega) \leq y) \to P(\omega : X_\infty(t,\omega) \leq y). \quad (9.27)$$

Relations (9.25)-(9.27) yield

$$P(\omega : X_\infty(t,\omega) \leq y) \to P_\infty(\omega : W(t,\omega) \leq y). \quad (9.28)$$

Lemma 9.3 *If A is a Borel set in $\mathbb{R}$ with the Lebesgue measure zero ($l(A) = 0$), then*

$$P(\omega : l(t \in [0,T] : X_\infty(t,\omega) \in A) = 0) = 1. \quad (9.29)$$

Proof For fixed $\epsilon > 0$ we first apply Chebyshev's inequality and Fubini's theorem, then we use (9.28) to derive that

$$P(\omega : l(t \in [0,T] : X_\infty(t,\omega) \in A) > \epsilon) \leq \epsilon^{-1} \mathbb{E} \int_0^T I_{(t:X_\infty(t,\omega)\in A)} dt$$

$$= \epsilon^{-1} \int_0^T \mathbb{E} I_{(t:X_\infty(t,\omega)\in A)} dt$$

$$= \epsilon^{-1} \int_0^T \left[\int_\Omega I_{(t:X_\infty(t,\omega)\in A)} dP \right] dt$$

$$= \epsilon^{-1} \int_0^T \left[\int_\Omega I_{(t:W(t,\omega)\in A)} dP_\infty \right] dt$$

$$= \epsilon^{-1} \int_0^T P_\infty(\omega : W(t,\omega) \in A) dt = 0.$$

Since $\epsilon > 0$ is arbitrary, we obtain (9.29). □

Let us define now the following process

$$X(t,\omega) = \int_0^t b(X_\infty(s,\omega))ds + W(t,\omega)$$

and consider the difference

$$X_n(t,\omega) - X(t,\omega) = \int_0^t b_n(X_n(s,\omega))ds - \int_0^t b(X_\infty(s,\omega))ds.$$

Applying Lemma 9.3 with $A = \{x_1, \ldots, x_N\}$ we find for almost all $t \in [0,T]$ w.r.to the Lebesgue measure and almost all $\omega \in \Omega$ that

$$\lim_{n\to\infty} b_n(X_n(t,\omega)) = b(X_\infty(t,\omega)).$$

Further, we derive from here by using the Lebesgue dominated convergence theorem that

$$\int_0^t b_n(X_n(s,\omega))ds \xrightarrow[n\to\infty]{} \int_0^t b(X_\infty(s,\omega))ds,$$

and, hence, for $t \in [0,T]$ and P-a.s.

$$X(t,\omega) = X_\infty(t,\omega),$$

which completes the proof of Theorem 9.3.

9.2 Diffusion processes and their connection with SDEs and PDEs

A general conception of a *Markov process* $(X_t)_{t\geq 0}$ on given stochastic basis $(\Omega, \mathcal{F}, (\mathcal{F}_t)_{t\geq 0}, P)$ can be described as follows. The characteristic (Markov) property of such adapted process means that

$$P(\omega : X(t, \omega) \in B | \mathcal{F}_s) = P(\omega : X(t, \omega) \in B | X(s, \omega)) \quad (9.30)$$

for every $B \in \mathcal{B}(\mathbb{R})$ and all $s < t$.

It is convenient to treat (9.30) with the help of a *transition probability function* $P(s, x, t, B)$ satisfying conditions:
(1) It is a measurable function of $x \in \mathbb{R}$,
(2) It is a probability measure on $(\mathbb{R}, \mathcal{B}(\mathbb{R}))$ such that $P(s, x, s, B) = I_B(x)$,
(3)

$$P(s, x, u, B) = \int P(s, x, t, dy) P(t, y, u, B). \quad (9.31)$$

for $x \in \mathbb{R}, B \in \mathcal{B}(\mathbb{R}), 0 \le s < t < u$.

The equation (9.31) is called the *Kolmogorov-Chapman* equation, and the process $(X_t)_{t \ge 0}$ is a *Markov random function* (in a restricted sense). Substituting x by $X(s, \omega)$ in (9.30) and using properties of conditional expectations we get

$$P(s, X(s, \omega), u, B) = \mathbf{E}(P(t, X(t, \omega), u, B) | \mathcal{F}_s) = \int P(t, y, u, B) P(s, X(s, \omega), t, dy)$$

which coincides with (9.31) under $x = X(s, \omega)$

We know that many properties of stochastic processes can be described using their finite-dimensional distributions. Let us derive the formula of a finite-dimensional distribution of $(X(t))$ in terms of its transition probability function. Putting $I_k = I_{B_k}(X(t_k, \omega)), 0 \le t_1 < t_2 < \ldots < t_k < \ldots t_n, B_k \in \mathcal{B}(\mathbb{R})$ we obtain by backward integration:

$$P(\omega : X(t_1, \omega) \in B_1, \ldots, X(t_n, \omega) \in B_n) = \mathbf{E} I_1 \ldots I_{n-1} \mathbf{E}(I_n | \mathcal{F}_{t_{n-1}})$$

$$= \mathbf{E} I_1 \ldots I_{n-1} \int_{B_n} P(t_{n-1}, X(t_{n-1}, \omega), t_n, dy_n)$$

$$= \mathbf{E} I_1 \ldots I_{n-1} \int_{B_{n-1}} P(t_{n-2}, X(t_{n-2}, \omega), t_{n-1}, dy_{n-1})$$

$$\times \int_{B_n} P(t_{n-1}, X(t_{n-1}, \omega), t_n, dy_n)$$

$$= \mathbf{E} I_1 \int_{B_2} P(t_1, X(t_1, \omega), t_2, dy_2) \times \ldots$$

$$\times \int_{B_n} P(t_{n-1}, X(t_{n-1}, \omega), t_n, dy_n). \quad (9.32)$$

In the expression above we can treat

$$\int_{B_2} P(t_1, X(t_1, \omega), t_2, dy_2) \ldots \int_{B_n} P(t_{n-1}, X(t_{n-1}, \omega), t_n, dy_n) \quad (9.33)$$

9.2 Diffusion processes and their connection with SDEs and PDEs

as a conditional expectation

$$P(\omega : X(t_2, \omega)) \in B_2, \ldots, X(t_n, \omega) \in B_n | X(t_1, \omega) = x_1).$$

The expressions (9.32)-(9.33) present a consistent system of finite-dimensional distributions of the process $X(t, \omega)$ for fixed t_1 and x_1. So, to provide a full description of this system it is reasonable to have a family of probability spaces $(\Omega, \mathcal{F}, P_{t,x})$. So, in case (9.32)-(9.33) we arrive to the probabilities $P_{t,x}(\omega : X(t_2, \omega) \in B_2, \ldots, X(t_n, \omega) \in B_n)$ expressed with the help of $P(s, x, t, B)$. We also can see that the process $X(t, \omega)$ can be started at each $t \in \mathbb{R}_+$ from each point $x \in \mathbb{R}$. The process $(X(t, \omega))$ in such a framework is called a *Markov family*, but we will call it also a *Markov process*. All these arguments speak us that in the theory of Markov processes it is necessary to describe the class of transition probability functions.

Before further steps in developments of this theory we give some examples and problems.

First of all, $(W_t)_{t \geq 0}$ presents an example of a Markov process, because its finite-dimensional distributions were given in its definition via transition probability function.

Problem 9.2 Prove that $(|W_t|)_{t \geq 0}$ is a Markov process.

Hint. Its transition function:

$$P(s, x, t, B) = \frac{1}{\sqrt{2\pi(t-s)}} \int_B \left[e^{-\frac{(y-x)^2}{2(t-s)}} + e^{-\frac{(y+x)^2}{2(t-s)}} \right] dy$$

Example 9.4 Let $(Y_n)_{n=1,2,\ldots}$ be a sequence of independent random variables with a positive density $p(x)$. Define $X_0 = 0$, $X_1 = Y_1^+$, $X_2 = (Y_1 + Y_2)^+, \ldots$ The process $(X_n)_{n=0,1,\ldots}$ is non-Markov. To show it we take $x > 0$ and consider the next conditional probability

$$P(\omega : X_3 > 0 | X_2 = 0, X_1 = x) = \int_{-\infty}^0 \int_0^\infty p(y-x) p(z-y) dy dz.$$

This probability depends on x, and therefore, the process (X_n) can not be Markov.

Problem 9.3 Let $(Y_n)_{n=1,2,\ldots}$ be a sequence of independent r.v.'s with the density $p(x) > 0$, $x \in \mathbb{R}$. Define the process $(Y_0 = 0)$

$$X_t = (Y_1 + \ldots + Y_k)(t - k) + (Y_1 + \ldots + Y_{k+1})(k + 1 - t)$$

for $k \leq t \leq k + 1$, $t \geq 0$. Prove that such a process is not Markov.

Remark 9.4 The transition probability function $P(s, x, t, B)$ is *homogeneous*, if $P(s + h, x, t + h, B) = P(s, x, t, B)$ for any positive h. Hence, $P(\cdot, \cdot, \cdot, \cdot)$ depends on the difference $(t - s)$ only and one can use another transition probability function $P(t, x, B)$ for which the Kolmogorov-Chapman equation has the form

$$P(t+s, x, B) = \int_{\mathbb{R}} P(t, x, dy) P(s, y, B).$$

Definition 9.1 Markov process $X(t, \omega)$ with its transition probability function $P(s, x, t, B)$ satisfying the properties (1)-(3) is called a *diffusion process*, if there exist measurable functions $b(t, x)$ and $\sigma^2(t, x), (t,) \in \mathbb{R}_+ \times \mathbb{R}$ such that for $\epsilon > 0$ and $V_\epsilon = \{y \in \mathbb{R} : |y - x| > \epsilon\}$

$$\lim_{h \to 0} \frac{1}{h} P(t, x, t + h, V_\epsilon) = 0, \tag{9.34}$$

$$\lim_{h \to 0} \frac{1}{h} \int_{|y-x| \le \epsilon} (y - x) P(t, x, t + h, dy) = b(t, x), \tag{9.35}$$

$$\lim_{h \to 0} \frac{1}{h} \int_{|y-x| \le \epsilon} (y - x)^2 P(t, x, t + h, dy) = \sigma^2(t, x). \tag{9.36}$$

Functions $b(t, x)$ and $\sigma^2(t, x)$ are called *drift* and *diffusion coefficients*.

It turns out relations (9.34)-(9.36) admit an adequate charazterization with the help of the following differential operator (generator of $X(t, \omega)$):

$$L_t \phi(x) = a(t, x) \frac{\partial \phi(x)}{\partial x} + \frac{1}{2} \sigma(t, x) \frac{\partial^2 \phi(x)}{\partial x^2}. \tag{9.37}$$

in class of twice continuously differentiable functions ϕ.

Lemma 9.4 *For all bounded functions ϕ described above*

$$\int_{\mathbb{R}} |\phi(y) - \phi(x)| P(t, x, t + h, dy) = h L_t \phi(x) + o(h) \tag{9.38}$$

is fulfilled $\iff$ (9.34)-(9.36).

Proof To prove (9.38), we decompose function ϕ at point x : for $|y - x| < \epsilon$

$$\int_{\mathbb{R}} |\phi(y) - \phi(x)| P(t, x, t + h, dy) = \alpha_\epsilon(h) + h L_t \phi(x) + h \gamma_\epsilon,$$

where L_t is defined by (9.37), $\alpha_\epsilon(h) h^{-1} \to 0$, $\epsilon > 0$, and $\gamma_\epsilon \to 0$ as $\epsilon \to 0$.

Hence, we get (9.38). To prove the direct implication we take the function $\phi_{x_0}(x) = 1 - \exp\left(-|x - x_0|^4\right)$ and find that $\frac{\partial}{\partial x} \phi_{x_0}(x)|_{x=x_0} = 0$ and $\frac{\partial^2}{\partial x^2} \phi_{x_0}(x)|_{x=x_0} = 0$. It implies (9.34).

Now we take function $\phi_{x_0}(y) = y - x_0$ for $|y - x_0| < 1$ extending this function to be a bounded twice continuously differentiable function. For such a function and $0 < \epsilon < 1$ we get

$$\int_{|y-x_0| \le \epsilon} (y - x_0) P(t, x_0, t + h, dy) = a(t, x) h + o(h) \tag{9.39}$$

9.2 Diffusion processes and their connection with SDEs and PDEs

and hence (9.35). To prove (9.36) we take a square of function in the previous construction (9.39). □

Let us find differential equations for transition probabilities of diffusion processes, known as the *Kolmogorov backward and forward equations*.

For some bounded continuous function $\phi(x)$ and $T > 0$ we define the following function

$$u_T(t, x) = \int_{\mathbb{R}} \phi(y) P(t, x, T, dy).$$

Theorem 9.4 *Assume that (9.34)-(9.36) are uniformly satisfied on bounded time intervals with continuous functions $b(t, x)$ and $\sigma^2(t, x)$. Let function $u_T(t, x)$ be (t, x)-continuous together with derivatives $\frac{\partial u_T}{\partial x}$ and $\frac{\partial^2 u_T}{\partial x^2}$. Then $u_T(t, x)$ has the derivative w.r.to t and satisfies the equation*

$$\frac{\partial u_T(t, x)}{\partial t} + L_t u_T(t, x) = 0, \quad (9.40)$$

with the boundary condition $u_T(T, x) = \phi(x)$.

Proof The boundary condition is fulfilled due to (9.34). Further, for $h > 0$ we have by Lemma 8.4 that

$$u_T(t - h, x) = \int_{\mathbb{R}} u_T(t, y) P(t - h, x, t, dy)$$

$$= u_T(t, x) + \int_{\mathbb{R}} [u_T(t, y) - u_T(t, x)] P(t - h, x, t, dy)$$

$$= h L_{t-h} u_T(t, x) + u_T(t, x) + o(h),$$

whence

$$\frac{1}{h} [u_T(t - h, x) - u_T(t, x)] = L_{t-h} u_T(t, x) + \frac{o(h)}{h}.$$

This implies (9.40). □

In fact, the equation (9.40) can be called as the Kolmogorov backward equation. But usually it is formulated as the equation for the *density* $p(s, z, t, y)$ of the transition probability function $P(s, x, t, B)$:

$$p(s, x, t, y) dy = P(s, x, t, dy).$$

So, we arrive to the *backward Kolmogorov equation* for $p(s, x, t, y)$:

$$\frac{\partial p(s, x, t, y)}{\partial s} + b(s, x,)\frac{\partial p(s, x, t, y)}{\partial x} + \frac{\sigma^2(s, x)}{2} \frac{\partial^2 p(s, x, t, y)}{\partial x^2} = 0. \quad (9.41)$$

On the other hand, one can write also the *forward Kolmogorov equation* or the *Fokker-Planck equation:*

$$\frac{\partial p(s, x, t, y)}{\partial t} + \frac{\partial}{\partial y}(b(t, y) p(s, x, t, y)) - \frac{\sigma^2}{2\partial y^2}\left(\sigma^2(t, y) p(s, x, t, y)\right) = 0. \quad (9.42)$$

So, the same function $p(s, x, t, y)$ plays the role a *fundamental solution* for two differential equations:
(9.41) as the function of (s, x) and
(9.42) as the function of (t, y).

Let us come back to the stochastic differential equation (9.1) and its solution $X(t)$ derived under the Lipschitz conditions in Theorem 9.1. Applying the same reasoning as in Theorem 9.1 we can prove that there exists a unique solution $X_{s,x}(t)$ of the equation $(t \geq s)$:

$$X_{s,x}(t) = x + \int_s^t b(u, X_{s,x}(u))du + \int_s^t \sigma(u, X_{s,x}(u))dW_u. \qquad (9.43)$$

Putting for $B \in \mathcal{B}(\mathbb{R})$
$$P(s, x, t, B) = P(X_{s,x}(t) \in B), \qquad (9.44)$$

we arrive to conclusion that the solution $X(t)$ of (9.1) is a *Markov process with the transition probability function* (9.44).

To prove this claim we note that $X_{s, X(s)}(t)$ is a solution of (9.43). We just note from (9.1) and Remark 9.1 that

$$X(t, \omega) - X(s, \omega) = \int_s^t b(u, X(u, \omega))du + \int_s^t \sigma(u, X(u, \omega))dW_u.$$

Thus, $X(t, \omega) = X_{s, X(s,\omega)}(t)$. Further, the σ-algebra $\mathcal{F}_s$ is generated by $W_u, u \leq s$. Let us take arbitrary bounded $\mathcal{F}_s$-measurable r.v. θ. Then for any bounded function $g(x)$ we obtain that

$$\begin{aligned}\mathbf{E}\theta g(X(t, \omega)) &= \mathbf{E}\theta g\left(X_{s, X(s,\omega)}(t)\right) \\ &= \mathbf{E}\theta \mathbf{E}\left(g\left(X_{s,x}(t)\right)\right)_{x=X(s,\omega)} \\ &= \mathbf{E}\theta \left(\int_{\mathbb{R}} g(y) P(s, x, t, dy)\right)_{x=X(s,\omega)} \\ &= \mathbf{E}\theta \int_{\mathbb{R}} g(y) P(s, X(s, \omega), t, dy),\end{aligned}$$

where we used the independence of $X_{s,x}(t)$ and $\mathcal{F}_s$ because $X_{s,x}(t)$ depends on $W(u) - W(s)$ for $u \geq s$ only. Hence,

$$\mathbf{E}(g(X(t, \omega))|\mathcal{F}_s) = \int_{\mathbb{R}} g(y) P(s, X(s, \omega), t, dy),$$

and we get the claim.

Remark 9.5 As a methodological result one can conclude that diffusion processes are described in two ways: as a class of Markov processes with transition probabilities determined by the drift and diffusion coefficients and as solutions of SDEs. Let us discuss other interesting connections. We consider a stochastic differential equation

9.2 Diffusion processes and their connection with SDEs and PDEs

with coefficients $b = b(t, x)$ and $\sigma = \sigma(t, x)$:

$$dX_t = b(t, X_t)dt + \sigma(t, X_t)dW_t.$$

As before, we define the following differential operator based on functions b and σ:

$$L\phi = \frac{1}{2}\sigma^2\frac{\partial^2\phi}{\partial x^2} + b\frac{\partial\phi}{\partial x}.$$

Having a smooth function $F \in C^{1,2}$ we can transform (X_t) with the help of the Ito formula as follows

$$F(X_t) = F(X_0) + \int_0^t \left[\frac{\partial F}{\partial s} + LF(X_s)\right] ds + \int_0^t \sigma(s, X_s)\frac{\partial F}{\partial x}(X_s)dW_s. \quad (9.45)$$

Equality (9.45) is a smooth transformation of the diffusion process (X_t). Such a problem was first formulated and solved by Kolmogorov in 1931: a new process $Y_t = F(X_t)$ will be again a diffusion process with the drift $b_Y = \frac{\partial F}{\partial t} + LF(X_t)$ and the diffusion coefficient $\sigma_Y = \sigma\frac{\partial F}{\partial x}(X_t)$. Putting together both these formulas we arrive to the Ito formula in the form (9.45). That's why the formula (9.45) can be also called the *Kolmogorov-Ito formula*.

Remark 9.6 We can also put the Cauchy problem for the parabolic differential equation: find a smooth function $v = v(t, x) : [0, T] \times \mathbb{R} \to \mathbb{R}$ such that

$$\frac{\partial v}{\partial t} + Lv = 0, \quad (t, x) \in [0, T] \times \mathbb{R}$$

with $v(T, x) = f(x)$, $x \in \mathbb{R}$.

This boundary value problem is solved in the theory of partial differential equations (PDEs) under wide conditions. It turns out one can write a probabilistic form of this solution using the theory of diffusion processes. We take a diffusion process (X_t) with a generator L and

$$v(t, x) = \mathbf{E}_{t,x} f(X_T), \quad X_t = x. \quad (9.46)$$

To get formula (9.46), called the *Feynman-Kac representation*, we apply the Ito formula to $v(t, X_t)$ and obtain that

$$dv(t, X_t) = \left[\frac{\partial v}{\partial t} + Lv\right] dt + \sigma\frac{\partial v}{\partial x}dW_t.$$

It is clear that $v(t, X_t)$ is a martingale, and therefore

$$v(t, X_t) = \mathbf{E}(f(X_T)|\mathcal{F}_t) = \mathbf{E}_{t,x} f(X_T)\big|_{x=X_t}.$$

Let us investigate absolute continuity of distributions of diffusion processes. We shall do it in the form of solutions of stochastic differential equations.

Assume $(X(t))_{t \leq T}$ is a continuous random process on a probabilistic space $(\Omega, \mathcal{F}, P)$. Denote $C[0, T]$ the space of continuous functions on $[0, T]$ and $\mathcal{B}^{[0,t]}$ is a σ-algebra in this space, generated by cylinders. We define a distribution of $(X(t))_{t \leq T}$ as the measure on $(C[0, T], \mathcal{B}^{[0,T]})$

$$\mu_X(B) = P\{\omega : X(\cdot, \omega) \in B\}.$$

It means that the measure μ_X is just an image of P under mapping

$$X(\cdot, \omega) : \Omega \to C[0, T].$$

Assume, there is a probability measure $\tilde{P} \ll P$, i.e. for $A \in \mathcal{F}$

$$\tilde{P}(A) = \int_A Z(\omega)dP,$$

where Z is a measurable non-negative random variable such that

$$\int_\Omega Z(\omega)dP = 1.$$

We can consider the process $(X_t)_{t \leq T}$ under a new measure $\tilde{P}$, and denote this process $(\tilde{X}_t)_{t \leq T}$. Then

$$\frac{d\mu_{\tilde{X}}}{d\mu_X}(X(\cdot, \omega)) = \mathbf{E}(Z(\omega)|\mathcal{F}^X), \qquad (9.47)$$

where $\mathcal{F}^X$ is a σ-algebra, generated by the process $(X_t)_{t \leq T}$. The equality (9.47) is almost obvious because for $A \in \mathcal{B}^{[0,T]}$ we have

$$\mu_{\tilde{X}}(A) = \tilde{P}(\omega : X(\cdot, \omega) \in A)$$
$$= \int_A Z(\omega)dP$$
$$= \mathbf{E}Z(\omega)I_A(X(\cdot, \omega))$$
$$= \mathbf{E}I_A(X(\cdot, \omega))\mathbf{E}(Z(\omega)|\mathcal{F}^X)$$
$$= \int_A \mathbf{E}(Z(\omega)|\mathcal{F}^X)\mu_X(dx).$$

Assume that two diffusion processes $(X_i(t))_{t \leq T}, i = 1, 2$, satisfy the following stochastic differential equations

$$\begin{aligned} dX_i(t) &= b_i(t, X_i(t))dt + \sigma(t, X_i(t))dW_t, \\ X_i(0) &= x \in \mathbb{R}. \end{aligned} \qquad (9.48)$$

Theorem 9.5 *Assume the coefficients of equations (9.48) are bounded and satisfy the Lipschitz conditions, and there exists a bounded continuous function $\theta = \theta(t, x)$ such that $\sigma(t, x) > 0$ and*

9.2 Diffusion processes and their connection with SDEs and PDEs

$$b_2(t, x) - b_1(t, x) = \theta(t, x)\sigma(t, x),$$

$(t, x) \in [0, T] \times \mathbb{R}$. Then $\mu_{X_2} \ll \mu_{X_1}$ and

$$\frac{d\mu_{X_2}}{d\mu_{X_1}}(X_1(\cdot, \omega)) = \exp\left\{\int_0^T \theta(s, X_1(s))dW_s - \frac{1}{2}\int_0^T \theta^2(s, X_1(s))ds\right\}. \quad (9.49)$$

Proof Obviously, we can apply here the Girsanov theorem with the measure $\tilde{P}$ and

$$Z(\omega) = \frac{d\tilde{P}}{dP} = \exp\left\{\int_0^T \theta(s, X_1)dW_s - \frac{1}{2}\int_0^T \theta^2(s, X_1)ds\right\},$$

and due to boundedness of coefficients $\mathbb{E}Z(\omega) = 1$.

Hence, $\tilde{P}(\Omega) = 1$ and the process

$$\tilde{W}(t) = W(t) - \int_0^t \theta(s, X_1(s))ds$$

is a Wiener process w.r. to $\tilde{P}$.

Let us rewrite the process $(X_1(t))_{t \leq T}$ as follows, using (9.5):

$$\begin{aligned}
X_1(t) - x &= \int_0^t b_1(s, X_1)ds + \int_0^t \sigma(s, X_1)dW_s \\
&= \int_0^t b_1(s, X_1)ds + \int_0^t \sigma(s, X_1)\left[d\tilde{W}_s + \theta(s, X_1)ds\right] \\
&= \int_0^t [b_1(s, X_1) + \theta(s, X_1)\sigma(s, X_1)]\,ds + \int_0^t \sigma(s, X_1)d\tilde{W}_s \\
&= \int_0^t b_2(s, X_1)ds + \int_0^t \sigma(s, X_1)d\tilde{W}_s.
\end{aligned} \quad (9.50)$$

Representation (9.50) certifies that $(X_1(t))_{t \leq T}$ coincides with the solution of the equation

$$d\tilde{X}_2(t) = b_2(t, \tilde{X}_2(t))dt + \sigma(t, \tilde{X}_2(t))d\tilde{W}_t$$

on the space $(\Omega, \mathcal{F}, \tilde{P})$. Hence, distributions $\mu_{\tilde{X}_2}$ and μ_{X_2} are the same, and applying (9.47) we obtain

$$\frac{d\mu_{X_2}}{d\mu_{X_1}}(X_1(\cdot, \omega)) = \mathbb{E}\left(Z(\omega)|\mathcal{F}^{X_1}\right). \quad (9.51)$$

Let us prove that $Z(\omega)$ is $\mathcal{F}^{X_1}$-measurable and therefore the formula (9.49) is fulfilled. To establish it we represent $\theta(t, x)$ as follows

$$\theta(t, x) = \frac{b_2(t, x) - b_1(t, x)}{\sigma(t, x)}$$

due to assumption $\sigma(t,x) > 0$ for $(t,x) \in [0,T] \times \mathbb{R}$. Further, for a subdivision $0 = s_0 < s_1 < \ldots < s_n = t$ we have (in the sense of convergence in probability) that

$$W_t = \lim_{n \to \infty} \sum_{i=0}^{n-1} \sigma^{-1}(s_i, X_1(s_i)) \left[X_1(s_{i+1}) - X_1(s_i) - \int_{s_i}^{s_{i+1}} b_1(s, X_1(s))ds \right]. \quad (9.52)$$

The relation (9.52) follows from observation that

$$\sum_{i=0}^{n-1} \sigma^{-1}(s_i, X_1(s_i)) \left[X_1(s_{i+1}) - X_1(s_i) - \int_{s_i}^{s_{i+1}} b_1(s, X_1(s))ds \right]$$

$$= \sum_{i=0}^{n-1} \sigma^{-1}(s_i, X_1(s_i))\sigma(s, X_1(s))dW_s,$$

which due to continuity of coefficients tends to W_t in probability as $\max_i \Delta s_i \to 0$ by properties of stochastic integrals. □

Example 9.5 Consider processes W_t, $X_t = 10 + W_t$ and $Y_t = 3W_t$, $t \in [0,1]$. Let us define a functional on $C[0,1]$, $f(x(\cdot)) = x(0)$. For the process X_t this functional takes value 10 with probability 1, but for other processes W_t and Y_t this functional takes value 0 with probability one. Hence, μ_X is singular w.r.to μ_W and μ_Y. We also note that μ_W and μ_Y are singular too. It follows from the fact that in probability $\sum_{i=1}^{n}(W_{t_i} - W_{t_{i-1}})^2 \to 1$ as the diameter of subdivision $0 = t_0 < t_1 < \ldots < t_n = 1$ tends to 1. A similar limit for (Y_t) will be equal to 9.

Remark 9.7 As it was noted in Remark 8.3 for Ito's processes, a similar theory of stochastic differential equations with a multidimensional Wiener process, a vector-valued coefficient b and a matrix-valued coefficient σ can be developed. Respectively, it is connected with diffusion processes with a vector-valued drift b, a matrix-valued diffusion $a = \sigma\sigma^*$ and a generator $Lu = \sum_i b^i \frac{\partial u}{\partial x^i} + \frac{1}{2}\sum_{i,j} a^{ij} \frac{\partial^2 u}{\partial x^i \partial x^j}$.

9.3 Applications to Mathematical Finance and Statistics of Random Processes

In financial context, a Wiener process was mathematically introduced and developed by L. Bachelier. His model of price evolution of stocks has the following simple form:

$$S_t = S_0 + \mu t + \sigma W_t, \quad t \leq T, \quad (9.53)$$

where $\mu \in \mathbb{R}$, $\sigma > 0$, and (W_t) is a Wiener process on given stochastic basis $(\Omega, \mathcal{F}, (\mathcal{F}_t^W), P)$.

Besides price dynamics of a risky asset (9.53) we assume for simplicity that evolution of non-risky asset (bank account) is trivial, i.e. $B_t = 1$.

9.3 Applications to Mathematical Finance and Statistics of Random Processes

One of the main subject of Mathematical Finance is option pricing. We consider only standard and the most exploited contracts which are called Call and Put options. These derivative securities (or simply, derivatives) give the right to the holder to buy (Call option) and to sell (Put option) a stock at maturity time T. To get such a derivative it is necessary to buy it by some price at time $t \le T$. Denote such prices for call and put options for $t = 0$ by $\mathbb{C}_T$ and $\mathbb{P}_T$, correspondently. It is convenient to identify these options with their pay-off functions at maturity time as $(S_T - K)^+$ and $(K - S_T)^+$. The problem is to find so-called *fair price* $\mathbb{C}_T(\mathbb{P}_T)$ in the beginning ($t = 0$) of the contract period. According to the theory of option pricing such a price $\mathbb{C}_T(\mathbb{P}_T)$ is calculated as $\mathbf{E}^*(S_T - K)^+$ (correspondingly, $\mathbf{E}^*(K - S_T)^+$), where $\mathbf{E}^*$ is the expectation with respect to a measure P^* such that the process $(S_t)_{t \le T}$ is a P^*-martingale. In case of the model (9.53) such *martingale* measure P^* is determined by the Girsanov exponent

$$Z_T^* = \exp\left\{-\frac{\mu}{\sigma}W_T - \frac{1}{2}\left(\frac{\mu}{\sigma}\right)^2 T\right\},$$

and according to the Girsanov theorem, the process

$$W_t^* = W_t + \frac{\mu}{\sigma}t$$

is a Wiener process w.r. to P^*. Hence, for distributions Law_{P^*} and Law_P w.r. to measures P^* and P relatively we have equality

$$Law_{P^*}(S_0 + \mu t + \sigma W_t; t \le T) = Law_P(S_0 + \sigma W_t; t \le T). \tag{9.54}$$

Theorem 9.6 *In the framework of the Bachelier model (9.53) the initial price of a call option is determined by the formula*

$$\mathbb{C}_T = (S_0 - K)\Phi\left(\frac{S_0 - K}{\sigma\sqrt{T}}\right) + \sigma\sqrt{T}\phi\left(\frac{S_0 - K}{\sigma\sqrt{T}}\right), \tag{9.55}$$

where $\Phi(x) = \int_{-\infty}^{x} \phi(y)dy$, $\phi(x) = \frac{1}{\sqrt{2\pi}}e^{-x^2/2}$.

In particular, for $S_0 = K$ we have $\mathbb{C}_T = \sigma\sqrt{\frac{T}{2\pi}}$.

To prove (9.55) we use (9.54) and self-similarity property of (W_t) and rewrite $\mathbb{C}_T$ as follows

$$\begin{aligned}\mathbb{C}_T &= \mathbf{E}^*(S_T - K)^+ = \mathbf{E}^*(S_0 + \mu T + \sigma W_T - K)^+ = \\ &= \mathbf{E}^*(S_0 + \sigma W_T - K)^+ = \mathbf{E}^*(S_0 + \sigma\sqrt{T}W_1 - K)^+ \\ &= \mathbf{E}^*(S_0 - K + \sigma\sqrt{T}\xi)^+\end{aligned} \tag{9.56}$$

where $W_1 = \xi \sim N(0, 1)$.

Denote $a = S_0 - K$ and $b = \sigma\sqrt{T}$ and obtain form (9.56) that

$$\begin{aligned}
\mathbb{C}_T &= \mathbf{E}(a + b\xi)^+ = \int_{-a/b}^{\infty} (a + bx)\phi(x)dx \\
&= a\Phi(a/b) + b\int_{-a/b}^{\infty} x\phi(x)dx \\
&= a\Phi(a/b) - b\int_{-a/b}^{\infty} d\phi(x) \\
&= a\Phi(a/b) + b\phi(a/b) \\
&= (S_0 - K)\Phi\left(\frac{S_0 - K}{\sigma\sqrt{T}}\right) + \sigma\sqrt{T}\phi\left(\frac{S_0 - K}{\sigma\sqrt{T}}\right).
\end{aligned}$$

Moreover, using the Markov property of (S_t) we can find the price of call option $\mathbb{C}(t, S_t)$ at any time $t \leq T$ taking conditional expected value of $(S_T - K)^+$ with respect to $\mathcal{F}_t$:

$$\begin{aligned}
\mathbb{C}(t, S_t) &= \mathbf{E}^*\left((S_T - K)^+ | \mathcal{F}_t\right) \\
&= \mathbf{E}^*\left((S_T - K)^+ | S_t\right) \\
&= \mathbf{E}_{t,x}(S_T - K)^+\Big|_{S_t = x} \\
&= \mathbf{E}(a_t + b_t\xi)^+ \\
&= a_t\Phi(a_t/b_t) + b_t\phi(a_t/b_t),
\end{aligned}$$

where $a_t = S_t - K$, $b = \sigma\sqrt{T - t}$.

Therefore,

$$\mathbb{C}(t, S_t) = (S_t - K)\Phi\left(\frac{S_t - K}{\sigma\sqrt{T - t}}\right) + \sigma\sqrt{T - t}\phi\left(\frac{S_t - K}{\sigma\sqrt{T - t}}\right). \tag{9.57}$$

Applying the Ito formula in (9.57) and using a martingale property of $\mathbf{E}^*((S_T - K)^+ | \mathcal{F}_t)$ we arrive to

$$d\mathbb{C}(t, S_t) = \frac{\partial \mathbb{C}}{\partial S_t}dS_t + \left(\frac{\partial \mathbb{C}}{\partial t} + \frac{1}{2}\sigma^2\frac{\partial^2 \mathbb{C}}{\partial S_t^2}\right)dt$$

and a PDE

$$\frac{\partial \mathbb{C}}{\partial t}(t, x) + \frac{1}{2}\sigma^2\frac{\partial^2 \mathbb{C}(t, x)}{\partial x^2} = 0. \tag{9.58}$$

with the boundary condition

$$\mathbb{C}(T, x) = (x - K)^+.$$

The equation (9.58) gives the opportunity to apply methods of PDEs in pricing of option.

9.3 Applications to Mathematical Finance and Statistics of Random Processes

Remark 9.8 As far as the put price $\mathbb{P}_T$ it follows from the *put-call parity*: for $r = 0$

$$\mathbb{C}_T = \mathbb{P}_T + S_0 - K. \tag{9.59}$$

The parity (9.59) follows from the next elementary equality

$$(x - K)^+ = (K - x) + x - K, \quad x, \ K \geq 0. \tag{9.60}$$

Putting S_T instead of x in (9.60), and taking expected value w.r.to P^* we arrive to (9.59).

The Bachelier model (9.53) has at least one disadvantage that prices can take negative values that contradicts their financial sense. To make a reasonable improvement of the model P. Samuelson (1965) proposed to transform (9.53) with the help of an exponential function. The resulting model

$$S_t = S_0 \exp\left((\mu - \sigma^2/2)t + \sigma W_t\right) \tag{9.61}$$

became the name of Geometric Brownian Motion (GBM).

Applying the Ito formula to (9.61) we get

$$dS_t = S_t(\mu dt + \sigma dW_t), \ S_0 > 0. \tag{9.62}$$

The model of financial market in the form (9.62) is called the Black-Scholes model. As in case of the Bachelier model, we assume here $B_t = 1$ for simplicity.

As in the case of the Bachelier model, we use the Girsanov theorem with the same Girsanov exponent Z_T^*, the martingale measure P^* and a Wiener process W_t^* w.r. to P^* we can conclude that

$$Law_{P^*}(\sigma W_t, t \leq T) = Law_{P^*}(\sigma W_t^*, t \leq T) = Law_P(\sigma W_t, \ t \leq T)$$

and, hence,

$$Law_{P^*}(S_t; t \leq T) = Law_P\left(S_0 e^{-\sigma^2/2t + \sigma W_t, t \leq T}\right).$$

It gives a possibility to calculate the price $\mathbb{C}_T$ of a Call option $(S_T - K)^+$ in the model (9.62) taking expected value w.r. to P^*:

$$\begin{aligned}\mathbb{C}_T &= \mathbf{E}^*(S_T - K)^+ = \mathbf{E}(S_0 e^{-\sigma^2/2T + \sigma W_T} - K)^+ \\ &= \mathbf{E}(S_0 e^{-\sigma^2/2T + \sigma\sqrt{T}W_1} - K)^+ \\ &= \mathbf{E}(ae^{b\xi - b^2/2} - K)^+ \\ &= a\Phi\left(\frac{\ln(\frac{a}{K}) + \frac{1}{2}b^2}{b}\right) - K\Phi\left(\frac{\ln(\frac{a}{K}) - \frac{1}{2}b^2}{b}\right),\end{aligned} \tag{9.63}$$

where $\xi \sim N(0, 1)$, $a = S_0$, $b = \sigma\sqrt{T}$.

The formula (9.63) is the famous formula of Black and Scholes for call option. A similar formula for put option is derived with the help of put-call parity.

Example 9.6 To recognize how close prices of call options in the model of Bachelier and the model of Black and Scholes we consider the simplest equations for them:

$$dS_t^B = S_0 \sigma dW_t,$$
$$dS_t^{BS} = S_t^{BS} \sigma dW_T, \ t \leq T, \sigma > 0.$$

We put $S_0 = K$, and in this case we compare prices $\mathbb{C}_T^B$ and $\mathbb{C}_T^{BS}$, for which

$$0 \leq \mathbb{C}_T^B - \mathbb{C}_T^{BS} \leq \frac{S_0}{12\sqrt{2\pi}} \sigma^3 T^{3/2} = O((\sigma\sqrt{T})^3). \tag{9.64}$$

To prove (9.64) we note the following inequalities: $e^y \geq 1 + y$ for all y, and, hence, $y^2/2 \geq 1 - e^{-y^2/2}$ for all y. Using these inequalities we obtain that

$$0 \leq \mathbb{C}_T^B - \mathbb{C}_T^{BS} = \left(\frac{S_0}{\sqrt{2\pi}} x - S_0 \left[\Phi\left(\frac{x}{2}\right) - \Phi\left(-\frac{x}{2}\right) \right] \right)_{x=\sigma\sqrt{T}}$$
$$= \frac{S_0}{\sqrt{2\pi}} \left[\int_{-x/2}^{x/2} \left(1 - e^{-y^2/2}\right) dy \right]_{x=\sigma\sqrt{T}}$$
$$\leq \frac{S_0}{\sqrt{2\pi}} \int_{-x/2}^{x/2} y^2/2 dy \Big|_{x=\sigma\sqrt{T}}$$
$$= \frac{S_0}{\sqrt{2\pi}} x^3/12 \Big|_{x=\sigma\sqrt{T}}$$
$$= \frac{S_0}{12\sqrt{2\pi}} \sigma^3 T^{3/2} = O((\sigma\sqrt{T})^3).$$

Assuming $\sigma\sqrt{T} \ll 1$ in (9.64) we can observe that call prices in both models are pretty close to each others. In particular, for $T = 1/2$ and $\sigma = 2.4\%$ we obtain $(\sigma\sqrt{T})^3 \approx 5 \cdot 10^{-7}$ and, therefore, $\frac{S_0}{12\sqrt{2\pi}} \cdot 5 \cdot 10^{-7} \approx 1.6 \cdot 10^{-8} S_0$.

Let us show how methods of stochastic processes work in solving of statistical problems.

Assume that the observation process has the following structure

$$X_t = \int_0^t f_s ds\theta + W_t, \tag{9.65}$$

where $(f_t)_{t\geq 0}$ is a progressively measurable process for which $P(\omega : \int_0^t f_s^2 ds < \infty) = 1$, $t \geq 0$, $\theta \in \mathbb{R}$ is an unknown parameter to be estimated based on observations of the process $(X_t)_{t\geq 0}$, and $(W_t)_{t\geq 0}$ is a Wiener process.

The model (9.65) is a continuous time version of of a regression model in discrete time. So, it is quite natural to use the *least squares estimates* for estimation of θ. In the case of model (9.65) such estimate has the following structure

9.3 Applications to Mathematical Finance and Statistics of Random Processes

$$\theta_t = \left(\int_0^t f_s^2 ds\right)^{-1} \int_0^t f_s dX_s.$$

We need to investigate some properties of estimates (9.3). We demonstrate it in the form of the following example.

Example 9.7 Assume that (a.s)

$$0 < c^2 \le f_t^2(\omega) \le C^2 < \infty, \tag{9.66}$$

for every $t > 0$. We can provide a "stochastic representation" of θ_t using (9.65) as follows

$$\theta_t = \left(\int_0^t f_s^2 ds\right)^{-1} \int_0^t f_s dX_S$$
$$= \left(\int_0^t f_s^2 ds\right)^{-1} \int_0^t f_s^2 ds\theta + \left(\int_0^t f_s^2 ds\right)^{-1} \int_0^t f_s dW_s,$$

and find that

$$\theta_t - \theta = \left(\int_0^t f_s^2 ds\right)^{-1} \int_0^t f_s dW_s. \tag{9.67}$$

Using representation (9.67), condition (9.66) and the Chebyshev inequality, we obtain for $\epsilon > 0$ that

$$P(\omega : |\theta_t - \theta| \ge \epsilon) \le \epsilon^{-2} \mathbb{E}\left(\left(\int_0^t f_s^2 ds\right)^{-1} \int_0^t f_s dW_s\right)^2$$
$$\le \epsilon^{-2} c^{-4} t^{-2} \mathbf{E}\left(\int_0^t f_s dW_s\right)^2$$
$$= \epsilon^{-2} c^{-4} t^{-2} \mathbf{E} \int_0^t f_s^2 ds$$
$$\le \epsilon^{-2} c^{-4} t^{-2} C^2 t$$
$$= \epsilon^{-2} \frac{C^2}{c^4} t^{-1} \to 0, \quad t \to \infty.$$

It means that θ_t is a consistent estimate of parameter θ.

Example 9.8 The least square estimate (9.3) is consistent, as was shown in the previous Example, but it does not speak us about the accuracy of these estimates. One can modify them with the help of a specially chosen stopping times to get an estimate with fixed accuracy. To do this we fix a positive number H and define

$$\tau_H = \inf\left(t : \int_0^t f_s^2 ds \ge H\right).$$

Assume $\int_0^t f_s^2 ds \to \infty$ (a.s.) as $t \to \infty$, to prove that

$$P(\omega : \tau_H(\omega) < \infty) = 1.$$

Further, define a *sequential least squares* estimate

$$\hat{\theta}_H = H^{-1} \int_0^{\tau_H} f_s dX_s. \qquad (9.68)$$

It follows from (9.68) that

$$\begin{aligned}
\mathbf{E}\hat{\theta}_H &= H^{-1} \mathbf{E} \int_0^{\tau_H} f_s dX_s \\
&= H^{-1} \mathbf{E} \int_0^{\tau_H} f_s^2 ds \theta + H^{-1} \mathbf{E} \int_0^{\tau_H} f_s dW_s \\
&= H^{-1} H \theta + H \cdot 0 = 0.
\end{aligned}$$

So, the estimate $\hat{\theta}_H$ is unbiased.

To estimate the accuracy of $\hat{\theta}_H$ we use its variance:

$$\begin{aligned}
\mathbf{Var}\hat{\theta}_H &= \mathbf{E}(\hat{\theta}_H - \theta)^2 \\
&= \mathbf{E}\left(H^{-1} \int_0^{\tau_H} f_s dX_s - \theta\right)^2 \\
&= H^{-2} \mathbf{E}\left(\int_0^{\tau_H} f_s dW_s\right)^2 \qquad (9.69) \\
&= H^{-2} \mathbf{E} \int_0^{\tau_H} f_s^2 ds \\
&= H^{-2} \cdot H = H^{-1}.
\end{aligned}$$

The relation (9.69) shows that accuracy of the estimate $\hat{\theta}_H$ can be controlled with the help of the level H.

9.4 Controlled diffusion processes and applications to option pricing

Suppose there is a family of diffusion processes $X_t = X_t^\alpha$ satisfying a stochastic differential equation

$$dX_t := dX_t^\alpha = b_\alpha(t, X_t^\alpha)dt + \sigma_\alpha(t, X_t^\alpha)dW_t, \quad X_0 = x. \qquad (9.70)$$

Here b_α and σ_α are functions satisfying some reasonable conditions for existence and uniqueness of solutions of (9.70). Parameter α is a *control process* adapted to

9.4 Controlled diffusion processes and applications to option pricing

filtration $(\mathcal{F}_t)_{t\geq 0}$. The equation (9.70) is also called a *stochastic control system*. We will call the process (X_t) a *controlled diffusion process*.

For estimating the quality of control α we introduce a function $f^\alpha(t,x)$, $(t,x) \in \mathbb{R}_+ \times \mathbb{R}$. The process α takes values in a domain $D \subseteq \mathbb{R}$. Function f^α can be interpreted as the density of the value flow. Then the total value on the interval $[0,t]$ will be identified with $\int_0^t f^\alpha(X_s^\alpha)ds$ which is assumed to be well defined.

Let us put the problem and provide some heuristic explanations of its solution for a time *homogeneous* stochastic control system (9.70). Denoting

$$v^\alpha(x) = \mathbf{E}_{0,x}\left[\int_0^\infty f^\alpha(X_s^\alpha)ds\right] = \mathbf{E}_x\left[\int_0^\infty f^\alpha(X_s^\alpha)ds\right],$$

we define the *optimal control* α^*, if

$$v(x) = \sup_\alpha v^\alpha(x) = v^{\alpha^*}(x), \quad x \in \mathbb{R}. \tag{9.71}$$

In the theory of controlled diffusion processes the following *Hamilton-Jacobi-Bellman principle of optimality* is exploited to determine $v(x)$:

$$v(x) = \sup_\alpha \mathbf{E}_x\left[\int_0^t f^\alpha(X_s^\alpha)ds + v(X_t^\alpha)\right]. \tag{9.72}$$

Let us explain a motivation for using (9.72). We rewrite the total value using strategy (control) α on $[0, \infty)$ as follows

$$\int_0^\infty f^\alpha(X_s^\alpha)ds = \int_0^t f^\alpha(X_s^\alpha)ds + \int_t^\infty f^\alpha(X_s^\alpha)ds. \tag{9.73}$$

If this strategy was used only up to moment t, then the first term in the right-hand side of (9.73) represents the value on the interval $[0,t]$. Suppose the controlled process $X_t^\alpha = X_t$ has the value $y = X_t$ at time t. If we wish to continue the control process after time t with the goal of maximization of the value over the whole time interval $[0, \infty)$, then we have to maximize $\mathbf{E}_y\left[\int_0^t f^\alpha(X_s^\alpha)ds\right]$, where α also denotes the continuation of the control process to $[t, \infty)$. Changing variable $s = t + u$, $u \geq 0$, and using independence and stationarity of increments of the Wiener process W, we obtain

$$\mathbf{E}_{X_t}\int_0^\infty f^\alpha(X_s^\alpha)ds = v^\alpha(X_t) \leq v(X_t).$$

Thus, a strategy that is optimal after time t, gives the average value such that

$$\mathbf{E}_x \int_0^t f^\alpha(X_s^\alpha)ds \geq v^\alpha(x). \tag{9.74}$$

One can choose α_s, $s \leq t$, so that the corresponding value is close enough to the average value. Therefore, taking supremum of both sides (9.74), we arrive to the HJB-principle of optimality (9.72), and we will call $v(x)$ the *value function*.

Moreover, one can rewrite (9.72) in a differential form if the value function is smooth enough. Applying the Ito formula, we obtain

$$v(X_t) = v(x) + \int_0^t \left[\frac{\partial v}{\partial x} b_\alpha(X_s) + \frac{1}{2}\sigma_\alpha^2 \frac{\partial^2 v}{\partial x^2}(X_s) \right] ds + \int_0^t \sigma_\alpha \frac{\partial v}{\partial x}(X_s) dW_s. \quad (9.75)$$

Since the last term in the right-hand side of (9.75) is a martingale, then we get from (9.72) that

$$v(x) = \sup_\alpha \mathbf{E}_x \left\{ \int_0^t \left[\frac{\partial v}{\partial x} b_\alpha(X_s) + \frac{\sigma_\alpha^2}{2} \frac{\partial^2 v}{\partial x^2}(X_s) + f^\alpha(X_s) \right] ds + v(x) \right\},$$

and hence,

$$\sup_\alpha \left[L_\alpha v(x) + f^\alpha(x) \right] = 0, \quad (9.76)$$

where $L_\alpha v = b_\alpha \frac{\partial v}{\partial x} + \frac{1}{2}\sigma_\alpha^2 \frac{\partial^2 v}{\partial x^2}$.

The equation (9.76) is usually referred to as *HJB-differential equation*.

Keeping in mind an adequate application of stochastic control theory in option pricing we would like reformulate it for the inhomogeneous case and for a finite interval time $[0, T]$. We consider the following stochastic control system

$$\begin{aligned} dX_s &= b_\alpha(s, X_s)ds + \sigma_\alpha(s, X_s)dW_s, \\ X_t &= x, \quad s \in [t, T], \end{aligned} \quad (9.77)$$

where α is a D-valued adapted process. Then the *optimal control problem* is to maximize (minimize) the value function

$$\begin{aligned} v^\alpha(t, x) &= \mathbf{E}_{t,x} \left[\int_t^T f^\alpha(s, X_s)ds + g(X_T) \right], \\ v(t, x) &= \sup_\alpha v^\alpha(t, x), \end{aligned} \quad (9.78)$$

where $X_s = X_s^\alpha$ satisfies (9.77) and function g determines the terminal value.

For system (9.77) with the value function (9.78) the corresponding HJB-equation is derived in the form

$$\begin{aligned} \frac{\partial v(t, x)}{\partial t} &+ \sup_\alpha L_\alpha v(t, x) = 0, \\ v(T, x) &= g(x), \quad x \in \mathbb{R}. \end{aligned} \quad (9.79)$$

Let us show how this mathematical technique works for option pricing. We start with the *Bachelier model with stochastic volatility* (with interest rate $r = 0$):

$$S_t = S_0 + \mu t + \sigma_t W_t, \quad t \in [0, T], \quad (9.80)$$

9.4 Controlled diffusion processes and applications to option pricing

where $S_0, \sigma_t^2 = \sigma^2 + (-1)^{N_t} \delta\sigma^2$, $\delta\sigma^2 < \sigma^2$, $(N_t)_{t \leq T}$ is a Poisson process with intensity $\lambda > 0$.

It is well-known that pricing of option with pay-off function g leads to the *interval of non-arbitrage prices* of option with end points

$$\mathbb{C}_* = \inf_{P^*} \mathbf{E}^* g(S_T) \text{ and } \mathbb{C}^* = \sup_{P^*} \mathbf{E}^* g(S_T), \qquad (9.81)$$

where P^* runs a family of martingale measures for the model (9.80).

Number $\mathbb{C}_*$ and $\mathbb{C}^*$ are called (initial) *lower* and *upper* price of option. For arbitrary time t before maturity date T such prices are determined as follows

$$\begin{aligned} v(t, S_t) &= \sup_{P^*} \mathbf{E}^*(g(S_T)|\mathcal{F}_t) \text{ and} \\ u(t, S_t) &= \inf_{P^*} \mathbf{E}^*(g(S_T)|\mathcal{F}_t). \end{aligned} \qquad (9.82)$$

Due to a Markov property of the process (X_t) from (9.80) and a parametrization of martingale measures $P^* = P^*(\alpha)$, $\alpha_t^2 = \sigma^2 + (-1)^{N_t} \delta\sigma^2$, formulas (9.81)-(9.82) can be rewritten as

$$\begin{aligned} v(t, S_t) &= \sup_{\alpha} \mathbf{E}^*(g(S_T)|S_t) \text{ and} \\ u(t, S_t) &= \inf_{\alpha} \mathbf{E}^*(g(S_T)|S_t). \end{aligned} \qquad (9.83)$$

Let us consider $v(t, S_t)$ only because the case of the lower price $u(t, S_t)$ (9.83) can be treated in the same way. Now we rewrite model (9.80) via $P^*(\alpha)$:

$$dS_t = \alpha(t, S_t)dW_t^*, \qquad (9.84)$$

where W^* is a Wiener process w.r. to P^*.

According the Ito formula we obtain

$$dv(t, S_t) = \left[\frac{\partial v(t, S_t)}{\partial t} + \frac{1}{2}\frac{\partial^2 v(t, S_t)}{\partial x^2}\alpha^2\right]dt + \alpha\frac{\partial v(t, S_t)}{\partial x}dW_t^*.$$

Hence, the equality (9.83) can be rewritten as

$$v(t, S_t) = \sup_{\alpha} \mathbf{E}_{t,S_t}\left[v(t, S_t) + \int_t^T \left[\frac{\partial v}{\partial s} + \frac{\alpha^2}{2}\frac{\partial^2 v}{\partial x^2}\right] + \int_t^T \alpha\frac{\partial v}{\partial x}dW_s^*\right]. \qquad (9.85)$$

Using the martingale property of the last term of (9.85) we arrive to the equation

$$0 = \sup_{\alpha} \mathbf{E}_{t,S_t}^*\left[\int_t^T \left[\frac{\partial v}{\partial s} + \frac{\alpha^2}{2}\frac{\partial^2 v}{\partial x^2}\right]ds\right] \qquad (9.86)$$

Divide both sides of (9.86) by $T - t$ and let $T \to t$ we get

$$0 = \sup_{\alpha}\left[\frac{\partial v(t, S_t)}{\partial t} + \frac{\alpha^2}{2}\frac{\partial^2 v(t, S_t)}{\partial x^2}\right]. \qquad (9.87)$$

Similarly, we can obtain that

$$0 = \inf_{\alpha}\left[\frac{\partial u(t, S_t)}{\partial t} + \frac{\alpha^2}{2}\frac{\partial^2 u(t, S_t)}{\partial x^2}\right]. \qquad (9.88)$$

Putting to (9.87)-(9.88) $\alpha^2 = \sigma^2 + (-1)^{N_t}\delta\sigma^2$, we determine $D = (\sigma^2 - \delta\sigma^2, \sigma^2 + \delta\sigma^2)$ and

$$0 = \frac{\partial v(t, S_t)}{\partial t} + \frac{1}{2}\frac{\partial^2 v(t, S_t)}{\partial x^2}\left(\sigma^2 + \operatorname{sgn}\left(\frac{\partial^2 v}{\partial x^2}(t, S_t)\right)\delta\sigma^2\right)$$

$$0 = \frac{\partial u(t, S_t)}{\partial t} + \frac{1}{2}\frac{\partial^2 u(t, S_t)}{\partial x^2}\left(\sigma^2 - \operatorname{sgn}\left(\frac{\partial^2 u}{\partial x^2}(t, S_t)\right)\delta\sigma^2\right).$$

So, we arrive to the following theorem of pricing of a call option in the model (9.80).

Theorem 9.7 *The bounds of non-arbitrage prices of a call option for any $t \leq T$ are calculated as solutions of the HJB-equations:*

$$\frac{\partial v(t, S_t)}{\partial t} + \frac{1}{2}\sigma^2\frac{\partial^2 v(t, S_t)}{\partial x^2} + \frac{1}{2}\left|\frac{\partial^2 v}{\partial x^2}\right|\delta\sigma^2 = 0, \qquad (9.89)$$

$$v(T, x) = (x - K)^+;$$

$$\frac{\partial u(t, S_t)}{\partial t} + \frac{1}{2}\sigma^2\frac{\partial^2 u(t, S_t)}{\partial x^2} - \frac{1}{2}\left|\frac{\partial^2 u}{\partial x^2}\right|\delta\sigma^2 = 0, \qquad (9.90)$$

$$u(T, x) = (x - K)^+.$$

The next theorem shows how equations (9.89) and (9.90) can be solved approximately by means of the small perturbations method.

Theorem 9.8 *Assume that $\delta\sigma^2 \ll \sigma^2$, then the initial upper and lower prices of a call option with strike price K admit approximations given by the formulas*

$$\hat{C}^* = (S_0 - K)\Phi\left(\frac{S_0 - K}{\sigma\sqrt{T}}\right) + \sigma\sqrt{T}\phi\left(\frac{S_0 - K}{\sigma\sqrt{T}}\right) + \frac{\delta\sigma^2}{2\sigma}\sqrt{T}\phi\left(\frac{S_0 - K}{\sigma\sqrt{T}}\right), \qquad (9.91)$$

$$\hat{C}_* = (S_0 - K)\Phi\left(\frac{S_0 - K}{\sigma\sqrt{T}}\right) + \sigma\sqrt{T}\phi\left(\frac{S_0 - K}{\sigma\sqrt{T}}\right) - \frac{\delta\sigma^2}{2\sigma}\sqrt{T}\phi\left(\frac{S_0 - K}{\sigma\sqrt{T}}\right), \qquad (9.92)$$

where $\phi(x) = \frac{1}{\sqrt{2\pi}}e^{-x^2/2}$, $\Phi(x) = \int_{-\infty}^{x}\phi(y)dy$.

Proof Both formulas (9.91) and (9.92) are derived in the same way using the method of small perturbations from the PDEs theory. Deriving (9.91) only we represent $v(t, x)$ as follows

$$v(t, x) = v_0(t, x) + v_1(t, x)\delta\sigma^2 + v_2(t, x)(\delta\sigma^2)^2 + \ldots \qquad (9.93)$$

9.4 Controlled diffusion processes and applications to option pricing

Let us make a change of variable $\theta = \sigma^2(T - t)$ and obtain $v(t, x) = V(\theta, x)$ and $\frac{\partial v}{\partial t} = -\sigma^2 \frac{\partial V}{\partial \theta}$.

Hence, the HJB-equation (9.89) is transformed to the following one

$$\frac{\partial V}{\partial \theta} = \frac{1}{2}\frac{\partial^2 V}{\partial x^2} + \frac{\delta\sigma^2}{2\sigma^2}\left|\frac{\partial^2 V}{\partial x^2}\right|, \quad V(0, x) = (x - K)^+. \quad (9.94)$$

To solve (9.94) we rewrite (9.93) in new variables ignoring terms of the order greater one:

$$V(\theta, x) = V_0(\theta, x) + V_1(\theta, x)\delta\sigma^2 + o(\delta\sigma^2).. \quad (9.95)$$

Plugging (9.95) to the equation (9.94) we get equations for V_0 and V_1:

$$\begin{aligned}\frac{\partial V_0}{\partial \theta} &= \frac{1}{2}\frac{\partial^2 V_0}{\partial x^2}, \quad V_0(0, x) = (x - K)^+, \\ \frac{\partial V_1}{\partial \theta} &= \frac{1}{2}\frac{\partial^2 V_1}{\partial x^2} + \frac{1}{2\sigma^2}\left|\frac{\partial^2 V_0}{\partial x^2}\right|, \quad V_1(0, x) = 0.\end{aligned} \quad (9.96)$$

The first equation in (9.96) is exactly the Bachelier equation, and we know its solution (see, for example, (9.57)). Moreover, we have for derivatives of V_0 that

$$\begin{aligned}\frac{\partial V_0}{\partial x} &= \Phi\left(\frac{x-K}{\sqrt{\theta}}\right) + (x-K)\phi\left(\frac{x-K}{\sqrt{\theta}}\right)\frac{1}{\sqrt{\theta}} \\ &\quad + (x-K)\phi\left(\frac{x-K}{\sqrt{\theta}}\right)\frac{-x+K}{\sqrt{\theta}} = \Phi\left(\frac{x-K}{\sqrt{\theta}}\right),\end{aligned}$$

$$\frac{\partial^2 V_0}{\partial x^2} = \phi\left(\frac{x-K}{\sqrt{\theta}}\right)\frac{1}{\sqrt{\theta}} \geq 0.$$

Therefore, the second equation in (9.96) is transformed to the next one

$$\frac{\partial V_1}{\partial \theta} = \frac{1}{2}\frac{\partial^2 V_1}{\partial x^2} + \frac{1}{2\sigma^2\sqrt{\theta}}\phi\left(\frac{x-K}{\sqrt{\theta}}\right),$$

for which we find that $V_1(\theta, x) = \frac{\sqrt{\theta}}{2\sigma^2}\phi(\theta, x)$.

Finally, we arrive to conclusion that

$$v(0, S_0) \approx V(\sigma^2 T, S_0) = (S_0 - K)\Phi\left(\frac{S_0 - K}{\sigma\sqrt{T}}\right) + \sigma\sqrt{T}\phi\left(\frac{S_0 - K}{\sigma\sqrt{T}}\right) + \frac{\delta\sigma^2}{2\sigma}\sqrt{T}\phi\left(\frac{S_0 - K}{\sigma\sqrt{T}}\right)$$

which proves the formula (9.91). □

Problem 9.4 Consider the Black-Scholes model with stochastic volatility and zero interest rate

$$\begin{aligned}dS_t &= S_t(\mu dt + \sigma_t dW_t), \quad S_0 > 0, \\ \sigma_t^2 &= s\sigma^2 + (-1)^{N_t}\delta\sigma^2, \quad \delta\sigma^2 < \sigma^2, \, t \in [0, T].\end{aligned} \quad (9.97)$$

Using a similar reasoning as in Theorem 9.7, derive the following HJB-equations for the upper and lower call option prices

$$\frac{\partial v}{\partial t} + \frac{1}{2}\sigma^2 x^2 \frac{\partial^2 v}{\partial x^2} + \frac{1}{2}\left|\frac{\partial^2 v}{\partial x^2}\right| x^2 \delta\sigma^2 = 0,$$

$$v(T, x) = (x - K)^+;$$

$$\frac{\partial u}{\partial t} + \frac{1}{2}\sigma^2 x^2 \frac{\partial^2 u}{\partial x^2} - \frac{1}{2}\left|\frac{\partial^2 u}{\partial x^2}\right| x^2 \delta\sigma^2 = 0,$$

$$u(T, x) = (x - K)^+.$$

Problem 9.5 Assuming in the model (9.97) that $\delta\sigma^2 \ll \sigma^2$, find the first order approximations for initial lower and upper prices $\mathbb{C}_*$ and $\mathbb{C}^*$ of call option:

$$\hat{\mathbb{C}}^* = S_0 \Phi\left(\frac{\ln S_0/K + \sigma^2 T/2}{\sigma\sqrt{T}}\right) - K\Phi\left(\frac{\ln S_0/K - \sigma^2 T/2}{\sigma\sqrt{T}}\right) + \frac{K\delta\sigma^2}{2\sigma^2}\sigma\sqrt{T}\phi\left(\frac{\ln S_0/K - \sigma^2 T/2}{\sigma\sqrt{T}}\right),$$

$$\hat{\mathbb{C}}_* = S_0 \Phi\left(\frac{\ln S_0/K + \sigma^2 T/2}{\sigma\sqrt{T}}\right) - K\Phi\left(\frac{\ln S_0/K - \sigma^2 T/2}{\sigma\sqrt{T}}\right) - \frac{K\delta\sigma^2}{2\sigma^2}\sigma\sqrt{T}\phi\left(\frac{\ln S_0/K - \sigma^2 T/2}{\sigma\sqrt{T}}\right).$$

Chapter 10
General theory of stochastic processes under "usual conditions"

Abstract Chapter 10 is devoted to a systematic exposition of a continuous time version of stochastic analysis under "usual conditions" with its standard notions like a stochastic basis, filtration, stopping times, random sets, predictable and optional sigma-algebras etc. It is shown how the discrete time martingale theory as well as a pure continuous time theory of diffusion processes are generalized for so-called cadlag processes. Using the predictable notion of a compensator the fundamental Doob-Meyer theorem is formulated for the class of sub- and supermartingales of class D. The full version of stochastic integration of predictable processes with respect to square-integrable martingale is developed. Moreover, different decompositions of such martingales are proved as well as the Kunuta-Watanabe inequality. It is shown how the theory can be extended with the help of localization procedures (local martingales, processes with locally integrable variation, semimartingales). The Ito formula is proved for semimartingales. SDEs with respect to semimartingales are studied including the existence and uniqueness of solutions of such equations with the Lipschitz coefficients (see [2], [8], [9], [16], [18], [20], [26], [33], [36], and [37]).

10.1 Basic elements of martingale theory

We will operate here with a complete probability space $(\Omega, \mathcal{F}, P)$ equipped with a non-decreasing family of σ-algebras $\mathcal{F}_t \subseteq \mathcal{F}$ satisfying the *"usual conditions"*:
1) $\mathcal{F}_s \subseteq \mathcal{F}_t$, $s \leq t$, $\mathcal{F}_t = \mathcal{F}_{t+} = \cap_{\epsilon>0} \mathcal{F}_{t+\epsilon}$;
2) $\mathcal{F}_t$ contains all the P-null sets of $\mathcal{F}$.

Let us call such a stochastic basis $(\Omega, \mathcal{F}, (\mathcal{F})_{t \geq 0}, P)$ *standard*.

In stochastic analysis, a special place is occupied by the set of *stopping times*. That is why we study this notion in a more general setting as before.

Definition 10.1 A non-negative random variable $\tau : \Omega \to \mathbb{R}_+ \cup \{\infty\}$ is a stopping time, if for any $t \geq 0$
$$\{\omega : \tau(\omega) \leq t\} \in \mathcal{F}_t.$$

We also define a σ-algebra

$$\mathcal{F}_\tau = \{A \in \mathcal{F}_\infty = \sigma\left(\cup_{t\geq 0}\mathcal{F}_t\right) : A \cap \{\tau \leq t\} \in \mathcal{F}_t\}$$

as a set of all events occurred before τ.

Let us formulate some properties of s.t.'s as the following problem.

Problem 10.1 1) If τ and σ are s.t.'s, then

$$\tau \vee \sigma = \max\{\tau, \sigma\} \text{ and}$$

$$\tau \wedge \sigma = \min\{\tau, \sigma\} \text{ are s.t.'s.}$$

2) If $(\tau_n)_{n=1,2,...}$ is a monotone sequence of s.t.'s, then

$$\tau = \lim_{n\to\infty} \tau_n \text{ is a s.t.}$$

3) If τ is a s.t., then $\mathcal{F}_\tau$ is a σ-algebra.
4) For two s.t.'s $\tau \leq \sigma$ we have $\mathcal{F}_\tau \subseteq \mathcal{F}_\sigma$.
5) Let τ be a s.t. and $A \in \mathcal{F}_\tau$, then

$$\tau_A = \begin{cases} \tau & \text{on } A, \\ \infty & \text{on } A^c \end{cases}$$

is a s.t.

6) Let τ be a s.t., then there exists a monotonic sequence of s.t.'s $\tau_n > \tau$ such that $\lim_{n\to\infty} \tau_n = \tau$ (a.s.).

Definition 10.2 A s.t. τ is called *predictable*, if there exists a non-decreasing sequence $(\tau_n)_{n=1,2,...}$ of s.t.'s such that

$$\lim_{n\to\infty} = \tau \text{ (a.s.), } \tau_n < \tau \text{ (a.s.) on } \{\tau > 0\}.$$

In this case we say that $(\tau_n)_{n=1,2,...}$ *announces* τ. Moreover, we denote $\mathcal{F}_{\tau-} = \sigma\left(\cup_{n=1}^\infty \mathcal{F}_{\tau_n}\right)$ as a collection of events *occurring strictly before* τ.

Problem 10.2 1) If $A \in \mathcal{F}_{\tau-}$, then τ_A is a predictable s.t.
2) If a.s. $\sigma < \tau$, then

$$\mathcal{F}_\sigma \subseteq \mathcal{F}_{\tau-} \subseteq \mathcal{F}_\tau.$$

Definition 10.3 For a s.t. τ we define a subset of $\mathbb{R}_+ \times \Omega$

$$[\![\tau]\!] = \{(t, \omega) : t = \tau(\omega) < \infty\}$$

as the *graph* of τ.

Definition 10.4 If the graph $[\![\tau]\!]$ can be embedded to a countable union of graphs of predictable s.t.'s, then a s.t. τ is *accessible*. It means that $\Omega = \cup_n A_n$, and on each

10.1 Basic elements of martingale theory

element A_n of such partition τ is announced by a sequence of s.t.'s $(\tau_{n,m})_{m=1,2,...}$. So, a s.t. τ will be *predictable* if the sequence $(\tau_{n,m})$ can be chosen without dependence on n.

If a s.t. τ such that $P(\omega : \tau = \sigma < \infty) = 0$ for every predictable s.t. σ, then τ is *totally inaccessible*.

Let us note that every s.t. τ can be decomposed as follows. There exists a unique (up to zero probability) set $A \in \mathcal{F}_\tau$ such that τ_A is accessible and τ_{A^c} is totally inaccessible, $A \subseteq \{\omega : \tau(\omega) < \infty\}$.

We need to connect these notion with the notion of a stochastic process. We understand a stochastic process $X = X_t(\omega) = X(t, \omega)$ as a mapping

$$X : \mathbb{R}_+ \times \Omega \to \mathbb{R}^d.$$

For simplicity we consider the case $d = 1$. Another stochastic process Y is a *modification* of X, if for arbitrary $t \geq 0$:

$$P\{\omega : X_t \neq Y_t\} = 0.$$

Two stochastic processes X and Y are *indistinguishable*, if

$$P\{\omega : X_t(\omega) = Y_t(\omega) \text{ for all } t\} = 1.$$

Example 10.1 To demonstrate a difference between the notion "modification" and "indistinguishability" we give a standard example. Take $X_t := 0$ and

$$Y_t = \begin{cases} 0, & t \neq \tau, \\ 1, & t = \tau, \end{cases}$$

where a r.v. τ is exponentially distributed with parameter $\lambda > 0$. Then for fixed t_0 we have $P(X_{t_0} = Y_{t_0}) = P(\tau \neq t_0) = 1$, but $P(X_t = Y_t \text{ for all } t) = 0$.

Definition 10.5 If X is a measurable mapping from $(\mathbb{R}_+ \times \Omega, \mathcal{B}(\mathbb{R}_+) \times \mathcal{F})$ to $(\mathbb{R}, \mathcal{B}(\mathbb{R}))$, then such a stochastic process is *measurable*.

If for $t \in \mathbb{R}_+$ a r.v. X_t is $\mathcal{F}_t$-measurable, then the process X is *adapted* (to filtration $(\mathcal{F}_t)$).

For each $t \geq 0$ we can consider a restriction of X to the set $[0, t] \times \Omega$. If such restriction is $\mathcal{B}([0, t]) \times \mathcal{F}_t$-measurable, then the process X is called *progressively measurable*. A family of sets $A \subseteq \mathbb{R}_+ \times \Omega$ such that the indicator process $X(t, \omega) = I_A(t, \omega)$ is progressively measurable is a *σ-algebra of progressively measurable sets* Π.

In case of a progressively measurable process X we have the following important property: X_τ is $\mathcal{F}_\tau$-measurable for any s.t. τ, where X_∞ must be $\mathcal{F}_\infty$-measurable .

Let us note that for any progressively measurable set A a random variable

$$D_A(\omega) = \inf\{t : (t, \omega) \in A\},$$

$D_A(\omega) = \infty$ if the set $\{\cdot\} = \emptyset$, is a stopping time which is called the *debut* of A.

One can note that every adapted and measurable process X admits a progressively measurable modification. It is convenient to operate with a stochastic process process X almost all trajectories of which are right-continuous and admit left-limits at each time. Such processes are called *cadlag*. It turns out such adapted cadlag process X is progressively measurable. This fact follows from the next standard considerations. For $t > 0, n = 1, 2, \ldots, i = 0, 1, \ldots, 2^n - 1, s \le t$ we put $X_0^{(n)}(\omega) = X_0(\omega)$

$$X_s^{(n)}(\omega) = X_{\frac{(i+1)t}{2^n}}(\omega) \text{ for } \frac{it}{2^n} < s \le \frac{i+1}{2^n}t. \tag{10.1}$$

According to (10.1) we get $X^{(n)}$ which is $\mathcal{B}(0, t) \times \mathcal{F}_t$-measurbale and due to the right-continuity $\lim_{n \to \infty} X_s^{(n)}(\omega) = X_s(\omega)$ for $(s, \omega) \in [0, t] \times \Omega$. Hence, X is progressively measurable.

Further, define the random variables

$$\tau_{1,0} = 0, \ \tau_{1,1} = \inf(t > 0 : |\Delta X_t| \ge 1), \ \tau_{1,2} = \inf(t > \tau_{1,1} : |\Delta X_t| \ge 1), \ldots$$

$$\tau_{k,0} = 0, \ldots, \tau_{k,n} = \inf\left(t > \tau_{k,n-1} : \frac{1}{k} \le |\Delta X_t| < \frac{1}{k-1}\right), \ldots, \tag{10.2}$$

where as usual $\Delta X_t = X_t - X_{t-}$.

These random variables are stopping times because (X_t) and (X_{t-}) are progressively measurable. Moreover, the set

$$U = \{(t, \omega) : \Delta X_t(\omega) \ne 0\} \subseteq \cup_{k \ge 1, n \ge 1} [\![\tau_{k,n}]\!]$$

and even can be embedded to the union of graphs $[\![\sigma_n]\!]$ and $[\![\tau_n]\!]$, where (σ_n) are predictable s.t.'s, (τ_n) are totally inaccessible s.t.'s such that $P(\sigma_n = \sigma_m < \infty) = 0$ and $P(\tau_n = \tau_m < \infty) = 0$, $n \ne m$.

Besides Π we need to introduce two additional σ-algebras on the space $[0, \infty) \times \Omega$. We shall do it with the help of the notion of *stochastic intervals*. These are examples of random set related to the stopping times σ and τ :

$$[\![\sigma, \tau[\![= \{(t, \omega) : \sigma(\omega) \le t < \tau(\omega)\},$$
$$[\![\sigma, \tau]\!] = \{(t, \omega) : \sigma(\omega) \le t \le \tau(\omega) < \infty\},$$

and so on, and also their graphs $[\![\sigma]\!] = [\![\sigma, \sigma]\!]$, $[\![\tau]\!] = [\![\tau, \tau]\!]$.

The filtration $(\mathcal{F}_t)_{t \ge 0}$ induces in $[0, \infty) \times \Omega$ besides Π the *predictable* and the *optional* σ-algebras $\mathcal{P}$ and O :

$$\mathcal{P} = \sigma\{[\![0_A]\!], [\![0, \tau]\!] : A \in \mathcal{F}_0, \tau \text{ is a s.t.}\}, \ [\![0_A]\!] = \{0\} \times A,$$
$$O = \sigma\{[\![0, \tau[\![: \tau \text{ is a s.t.}\}.$$

The processes which are measurable w.r. to $\mathcal{P}$ (correspondently, O) are called *predictable* (*optional*).

10.1 Basic elements of martingale theory

Problem 10.3 Let (τ_i) is a non-decreasing sequence of s.t.'s and bounded random functions ϕ_0^* and ϕ_i are $\mathcal{F}_0$ and $\mathcal{F}_{\tau_i}$-measurable correspondingly. Define a stochastic process

$$X_t(\omega)\phi_0^*(\omega)I_{\{0\}}(t) + \sum_{i=0}^{n-1} \phi_i(\omega)I_{\rrbracket\tau_i,\tau_{i+1}\rrbracket}(t,\omega). \tag{10.3}$$

Prove that $\mathcal{P}$ is generated by all processes of type (10.3).

Remark 10.1 One can prove that $\mathcal{P}$ is generated by processes of type (10.3) with *deterministic* $\tau_i = t_i$. Moreover, $\mathcal{P}$ is generated by all left-continuous (continuous) adapted processes.

It is clear that for the left-continuous process X and a predictable s.t. τ $X_\tau I_{\{\tau<\infty\}}$-$\mathcal{F}_{\tau-}$-measurable, and therefore such a statement will be true for all predictable processes and times.

Remark 10.2 If A is a predictable set ($A \in \mathcal{P}$), then its debut is a s.t. Moreover, D_A is predictable and $A \setminus \llbracket D_A, \infty \llbracket \in \mathcal{P}$, if $\llbracket D_A \rrbracket \subseteq A$.

Remark 10.3 As far as the times $\tau_{k,n}$ of the n-th jump of size $|\Delta X_t| \in \left[\frac{1}{k}, \frac{1}{k-1}\right)$, for an adapted predictable process X (see (10.2)) they are predictable, $U = \cup_{n\geq 1, k\geq 1}\llbracket \tau_{k,n} \rrbracket$, and $X_{\tau_{n,k}}$-$\mathcal{F}_{\tau_{n,k-}}$-measurable, according to Remark 10.1.

To go further, we will study two classes of stochastic processes: *processes with finite variation* and *martingales*.

Definition 10.6 1) The process $A = (A_t)_{t\geq 0}$ is *increasing*, if $A_0 = 0$ (a.s.), adapted and cadlag, $A_s \leq A_t$ (a.s.) for $s \leq t$.

2) The process $B = (B_t)_{t\geq 0}$ is the *process with finite variation*, if $B_0 = 0$ (a.s.), adapted and cadlag, for each $\omega \in \Omega$ the trajectory $B_\cdot(\omega)$ has finite variation on each compact interval.

Each process with finite variation can be decomposed into difference of two increasing processes and vice versa. For any bounded (non-negative) Borel function $f(s)$ one can define the Lebesgue-Stieltjes integral $\int_0^t f(s)dB_s(\omega)$. We denote $\int_0^t f(s)|dB_s|$ the integral of f under the variation of B, and $\int_0^t |dB_s|$ will be equal to the variation of B on $[0,t]$.

Definition 10.7 We call an increasing process A *integrable* if $\mathbf{E}\int_0^t dA_s < \infty$. Correspondently, a process B has *integrable variation* if $\mathbf{E}\int_0^\infty |dB_s| < \infty$. The corresponding classes of such processes we denote $\mathcal{A}^+$ and $\mathcal{A}$.

The process B with finite variation can be decomposed as follows

$$B_t = B_t^c + \sum_{s\leq t} \Delta B_s, \tag{10.4}$$

where B^c is a continuous process with finite variation and $\sum_{s\leq t}|\Delta B_s| \leq \int_0^t |dB_s| < \infty$.

Moreover, in decomposition (10.4) its discontinuous part $\sum_{s \leq t} \Delta B_s = \sum_{n \geq 1} \Delta B_{\tau_n} I_{\{\tau_n \leq t\}}$ with some sequence of s.t.'s $(\tau_n)_{n \geq 1}$. If the process B is predictable, then τ_n are predictable too, and $\Delta B_{\tau_n} - \mathcal{F}_{\tau_n-}$-measurable. Hence, any *predictable process with finite variation* admits the form (10.4) as

$$B_t = B_t^c + \sum_{n \geq 1} \phi_n I_{\{\tau_n \leq t\}}, \tag{10.5}$$

where $\phi_n - \mathcal{F}_{\tau_n-}$-measurable and $\sum_{n \geq 1} |\phi_n| I_{\{\tau_n \leq t\}}$ exists for all $t \in \mathbb{R}_+$.

In fact, if B satisfies (10.5) then it is a predictable process with finite variation.

Definition 10.8 A *martingale* is an adapted integrable process $M = (M_t)_{t \geq 0}$ such that

$$\mathbf{E}(M_t | \mathcal{F}_s) = M_s \ (a.s.) \ for \ all \ t \geq s \geq 0. \tag{10.6}$$

If in (10.6) the "=" is changed by "≥" (correspondently, "≤"), then we have a *submartingale (supermartingale)*.

Without loss of generality we always can count these processes *cadlag*, because under "usual conditions" every submartingale admits the right-continuous modification if and only if its expected value is right-continuous.

We can transform any martingale $(M_t)_{t \geq 0}$ to a submartingale $(\phi(M_t))_{t \geq 0}$ with the help of a convex function ϕ such that $\mathbf{E}|\phi(M_t)| < \infty$, $t \in \mathbb{R}_+$, using *Jensen's inequality*.

As in discrete time, we can define for given process $X = (X_t)_{t \geq 0}$ the following notions. Let $a < b$ be real numbers and F be a finite subset of $\mathbb{R}_+$. For each (restricted to F) trajectory of process $(X_t(\omega))_{t \geq 0}$ one can define the number of upcrossings $U_F(a, b, X.(\omega))$ of the interval $[a, b]$. To do it we take

$$\tau_1(\omega) = \inf(t \in F : X_t(\omega) \leq a) \text{ and for } j = 1, 2, \ldots$$
$$\sigma_j(\omega) = \inf(t \in F : t \geq \tau_j(\omega), X_t(\omega) > b),$$
$$\tau_{j+1}(\omega) = \inf(t \in F : t \geq \sigma_j(\omega), X_t(\omega) < a),$$

and define $U_F(a, b, X.(\omega))$ the largest j such that $\sigma_j(\omega) < \infty$.

For an infinite set $G \subseteq \mathbb{R}_+$ one can define

$$U_G(a, b, X.(\omega)) = \sup\{U_F(a, b, X.(\omega)) : F \subseteq G, \ F \text{ is finite}\}.$$

Let us formulate the following *Doob inequalities* for continuous time submartingales.

Theorem 10.1 *Assume $X = (X_t)_{t \geq 0}$ is a submartingale, $[t_1, t_2] \subseteq \mathbb{R}_+$, $\lambda > 0$. Then*

1) $P\left\{\omega : \sup_{t_1 \leq t \leq t_2} X_t \geq \lambda\right\} \leq \frac{\mathbf{E} X_{t_2}^+}{\lambda}$;

2) $P\left\{\omega : \inf_{t_1 \leq t \leq t_2} X_t \leq -\lambda\right\} \leq \frac{\mathbf{E} X_{t_2}^+ - \mathbf{E} X_{t_1}}{\lambda}$;

3) $\mathbf{E} U_{[t_1, t_2]}(a, b, X.(\omega)) \leq \frac{\mathbf{E} X_{t_2}^+ + |a|}{b-a}$;

4) $\mathbf{E}\left(\sup_{t_1 \leq t \leq t_2} X_t\right)^p \leq \left(\frac{p}{p-1}\right)^p \mathbf{E} X_{t_2}^p$, $p > 1$,

for a non-negative (X_t) wih $\mathbf{E} X_t^p < \infty$.

10.1 Basic elements of martingale theory

The *derivation* of inequalities of the above theorem is based on limiting arguments, right-continuity of X and discrete time versions of these results.

Let us pay attention to other important results of martingale theory.

Theorem 10.2 *If $X = (X_t)_{t \geq 0}$ is a submartingale with $\sup_{t \geq 0} EX_t^+ < \infty$, then there exists a r.v. $X_\infty - \mathcal{F}_\infty$-measurable and integrable such that $X_\infty(\omega) = \lim_{t \to \infty} X_t(\omega)$ (a.s.).*

Proof For $n = 1, 2, \ldots$ and $a < b \in \mathbb{R}$ we have from Theorem 9.1 that

$$EU_{[0,n]}(a, b, X.(\omega)) \leq \frac{EX_n^+ + |a|}{b - a}. \tag{10.7}$$

Taking the limit as $n \to \infty$ in (10.7) we get

$$EU_{[0,\infty[}(a, b, X.(\omega)) \leq \frac{\sup_t EX_n^+ + |a|}{b - a}. \tag{10.8}$$

Denote $A_{a,b} = \{\omega : U_{[0,\infty)}(a, b, X.(\omega)) = \infty\}$ and find from (10.8) that $P(A_{a,b}) = 0$. Hence, $P(A) = P(\cup_{a,b \in \mathbb{Q}} A_{a,b}) = 0$. The set A contains the set $\{\omega : \limsup_{t \to \infty} X_t(\omega) > \liminf_{t \to \infty} X_t(\omega)\}$.

Now, for $\omega \in \Omega \setminus A$ there exists

$$X_\infty(\omega) = \lim_{t \to \infty} X_t(\omega) \ (a.s.)$$

Due to $E|X_t| = 2EX_t^+ - EX_t \leq 2 \sup_t EX_t^+ - EX_0$ and Fatou's lemma we obtain that $E|X_\infty| < \infty$. □

Corollary 10.1 *Let $X = (X_t)_{t \geq 0}$ be a martingale. Then the following statements are equivalent*

1) *$(X_t)_{t \geq 0}$ is uniformly integrable;*
2) *X_t converges to X_∞ in L^1;*
3) *There exists $X_\infty \in L^1$ such that $X_t = E(X_\infty | \mathcal{F}_t)$.*

Corollary 10.2 *Let $X = (X_t)_{t \geq 0}$ be a nonnegative supermartingale. Then there exists an integrable r.v. $X_\infty = \lim_{t \to \infty} X_t$ (a.s.).*

Corollary 10.3 *Let $X = (X_t)_{t \in [0,\infty]}$ be a submartingale on extended real line $\mathbb{R}_+ \cup \{\infty\}$. Then for stopping times $\sigma \leq \tau$*

$$E(X_\tau | \mathcal{F}_\sigma) \geq X_\sigma (a.s.) \tag{10.9}$$

Proof Consider the following sequence of discrete stopping times

$$\sigma_n(\omega) = \begin{cases} \sigma(\omega), & \sigma = \infty, \\ \frac{kn}{2}, & \frac{k-1}{2^n} \leq \sigma(\omega) < \frac{k}{2^n}. \end{cases}$$

Similar sequence is constructed for τ, and by construction and conditions of the theorem $\sigma_n \leq \tau_n$, $n = 1, 2, \ldots$. Applying (10.9) with σ_n and τ_n, we derive (10.9) by taking the limit as $n \to \infty$. □

Corollary 10.4 *Let $X = (X_t)_{t \geq 0}$ be a submartingale and $\sigma \leq \tau$ be stopping times. Then*
1) $(X_{\tau \wedge t})_{t \geq 0}$ is a submartingale w.r.to $(\mathcal{F}_t)$;
2) $\mathbb{E}(X_{\tau \wedge t}|\mathcal{F}_\sigma) \geq X_{\sigma \wedge t}$ (a.s.) for all $t \geq 0$.

Corollary 10.5 *If $M = (M_t)_{t \geq 0}$ is a uniformly integrable martingale and τ is a predictable stopping time, then $\mathbf{E}(\Delta M_\tau | \mathcal{F}_{\tau-}) = 0$.*

To prove it we consider a sequence of stopping times $(\tau_n)_{n=1,2,\ldots}$ announcing τ. Applying Theorem 10.2 to this sequence we get $M_{\tau_n} = \mathbf{E}(M_\infty | \mathcal{F}_{\tau_n}) = \mathbf{E}(M_\tau | \mathcal{F}_{\tau_n})$. Taking the limit in this equality we obtain (a.s.)

$$M_{\tau-} = \lim_{n \to \infty} M_{\tau_n} = \lim_{n \to \infty} \mathbf{E}(M_\tau | \mathcal{F}_{\tau_n}) = \mathbf{E}(M_\tau | \mathcal{F}_{\tau-}).$$

Denote $\mathcal{M}$ the class of *uniformly integrable martingales*.

Assume $M \in \mathcal{M}$ and $A \in \mathcal{A}^+$, then the process $X = M - A$ is a supermartingale, satisfying the *condition (D)*, i.e. the class of random variables $\{X_\tau I_{\{\tau < \infty\}} : \tau$ is a s.t.$\}$ is uniformly integrable. Hence, such a construction leads to a supermartingale $X \in D$. The inverse statement is the *famous Meyer's theorem*:

Any supermartingale $X \in D$ admits the following *decomposition of Doob-Meyer*: for $t \geq 0$ (a.s.)

$$X_t = M_t - A_t, \tag{10.10}$$

where $M \in \mathcal{M}$ and A is a *predictable* process from $\mathcal{A}^+$. The decomposition (10.10) is unique in the class of predictable increasing processes.

Remark 10.4 If a supermartingale $X \in D$ and B is the predictable increasing process from the Doob-Meyer decomposition, then we can calculate the jumps of B as follows. First of all, the moments of jumps of B are predictable. For any such a moment τ we can apply Corollary 10.4 with the uniformly integrable martingale $X = M + B$ and get

$$0 = \mathbf{E}(\Delta M_\tau | \mathcal{F}_{\tau-}) = \mathbf{E}(\Delta X_\tau + \Delta B_\tau | \mathcal{F}_{\tau-}) = \mathbf{E}(\Delta X_\tau | \mathcal{F}_{\tau-}) + \Delta B_\tau,$$

and hence $\Delta B_\tau = -\mathbf{E}(\Delta X_\tau | \mathcal{F}_{\tau-})$.

Let us note that for every $A \in \mathcal{A}^+$ the process $(-A)$ is a supermartingale of class (D).

Applying to $(-A)$ its Doob-Meyer decomposition, we find a *unique predictable* process $B \in \mathcal{A}^+$ such that $B - A \in \mathcal{M}$.

Definition 10.9 *The process B is called the compensator of A.*

These considerations can be extended to the class $\mathcal{V}$ *of processes with integrable variation*. Let $A \in \mathcal{V}$, then there exists a *unique predictable* process $B \in \mathcal{V}$ with the property: $B - A \in \mathcal{M}$. We will call B the *compensator* of A.

Remark 10.5 a) If τ is a totally inaccessible stopping time and ϕ is an integrable $\mathcal{F}_\tau$-measurable function, then the *compensator* of $A_t = \phi I_{\{\tau \leq t\}}$ is a *continuous* process from class $\mathcal{V}$.

10.1 Basic elements of martingale theory

b) If τ is predictable and ϕ is an integrable $\mathcal{F}_\tau$-measurable function, then the compensator of $A_t = \phi I_{\{\tau \le t\}}$ is the process $B_t = \mathbf{E}(\phi|\mathcal{F}_{\tau-})I_{\{\tau \le t\}}$.

The rest of this Section is devoted to *square-integrable martingales*.

Definition 10.10 A martingale M is *square integrable* if $\sup_t \mathbf{E} M_t^2 < \infty$. The class of such martingales is denoted of $\mathcal{M}^2$.

We count some properties of such martingales in the following theorem.

Theorem 10.3 *Let $M \in \mathcal{M}^2$, then the next properties are fulfilled:*
(1) $(M_t^2)_{t \ge 0}$ is a submartingale;
(2) there exists $M_\infty = \lim_{t \to \infty} M_t$ (a.s.) and in L^2;
(3) $\mathbf{E} \sup_t M_t^2 \le 4 \sup_t \mathbf{E} M_t^2 = 4 \mathbf{E} M_\infty^2$;
(4) $\mathbf{E} \sum_t (\Delta M_t)^2 \le \liminf_{(t_1 < ... < t_n)} \mathbf{E} \sum (M_{t_i} - M_{t_{i-1}})^2 = \mathbf{E} M_\infty^2 < \infty$.

Proof Statements (1)-(3) come from Jensen's inequality and Theorem 10.1 and 10.2. The last one 4) follows from the Fatou lemma, where liminf is provided via directed sets if subdivisions $(t_1 < ... < t_n)$. □

Let us note that due to the Doob inequality M^2 is a submartingale from class (D), and hence, by the Doob-Meyer decomposition there exists a unique predictable $A \in \mathcal{A}^+$ such that $M^2 - A \in \mathcal{M}$.

The process A is denoted $\langle M, M \rangle$ and is called a *quadratic characteristic (compensator)* of $M \in \mathcal{M}^2$.

Remark 10.6 We can cover the jumps of $M \in \mathcal{M}^2$ by a sequence of predictable s.t.'s $(\tau_n)_{n=1,2,...}$ and a sequence of totally inaccessible s.t.'s $(\sigma_n)_{n=1,2,...}$.

For each $n = 1, 2, ...$ we define processes $C_t^n = \Delta M_{\tau_n} I_{\{\tau_n \le t\}}$ and $D_t^n = \Delta M_{\sigma_n} I_{\{\sigma_n \le t\}}$.

Due to Remark 10.5 in case C_t^n we have $\mathbf{E}(\Delta M_{\tau_n}|\mathcal{F}_{\tau_n-})I_{\{\tau_n \le t\}} = 0 \cdot I_{\{\tau_n \le t\}} = 0$, and hence, it is a square integrable martingale. Another process D_t^n has a compensator $\tilde{D}_t^n$ which is a continuous process with integrable variation. So, $\hat{D}_t^n = D_t^n - \tilde{D}_t^n \in \mathcal{M}$ and $\Delta \hat{D}_{\sigma_n}^n = \Delta M_{\sigma_n}$.

To investigate the structure of class $\mathcal{M}^2$ we need several auxiliary facts, which we formulate as Lemmas.

Lemma 10.1 *Let A and $B \in \mathcal{A}$, $A - B$ be a martingale and ϕ be a bounded (nonnegative) predictable function. Then*

$$\mathbf{E} \int_0^\infty \phi_s d(A - B)_s = 0. \qquad (10.11)$$

Proof It is clear that relation (10.11) is fulfilled for the step function (see (10.3) and Remark 10.1)

$$\phi_t = \phi(t) = \phi_0^* I_{\{0\}}^{(t)} + \sum_{i=0}^{n-1} \phi_i(\omega) I_{(t_i, t_{i+1}]}(t),$$

and extension to predictable bounded functions is straightforward. □

Lemma 10.2 *If $a : \mathbb{R}_+ \to \mathbb{R}_+$ is the right-continuous increasing function, $a(0) = 0$, and $c(s) = \inf(t : a(t) > s)$, then $c(s)$ is the increasing right-continuous satisfying the equality*

$$\int_0^\infty f(s)da(s) = \int_0^\infty f(c(s))I_{\{c(s)<\infty\}}ds \quad (10.12)$$

for any bounded (non-negative) measurable function f.

Again, the *proof* is standard. We check (10.12) for $f(s) = I_{(0,t]}(s)$ and so on.

Lemma 10.3 *For any bounded martingale M and process $A \in \mathcal{A}$ the following equality is true*

$$\mathbf{E}A_\infty M_\infty = \mathbf{E}\int_0^\infty M_s dA_s. \quad (10.13)$$

Proof Using (10.12) with A and c, we obtain (10.13) from the following chain of equalities

$$\mathbf{E}\int_0^\infty M_s dA_s = \mathbf{E}\int_0^\infty M_{c_s} I_{\{c_s<\infty\}}ds$$

$$= \int_0^\infty \mathbf{E}M_{c_s} I_{\{c_s<\infty\}}ds$$

$$= \int_0^\infty \mathbf{E}\left(M_\infty I_{\{c_s<\infty\}}\right) ds$$

$$= \mathbf{E}\int_0^\infty M_\infty I_{\{c_s<\infty\}}ds$$

$$= \mathbf{E}\int_0^\infty M_\infty dA_s$$

$$= \mathbf{E}M_\infty dA_\infty.$$

The claim of the next Lemma was proved before for discrete time martingales.

Lemma 10.4 *Let $(L_t)_{t\in[0,\infty]}$ be an uniformly integrable process, $L_0 = 0$. If $\mathbf{E}L_\sigma = 0$ for any stopping time σ, then L is a martingale.*

Proof We define a s.t.

$$\sigma = \begin{cases} t & \text{on } A, \\ \infty & \text{on } A^c, \end{cases}$$

where A is a fixed set from $\mathcal{F}_t$. Then we obtain that $\int_A L_t dP + \int_{A^c} L_\infty dP = 0$, which can be transformed for $\sigma = \infty$ to the equality $\int_A L_\infty dP + \int_{A^c} L_\infty dP$, i.e. (L_t) is a martingale. □

Lemma 10.5 *For $M \in \mathcal{M}^2$ and any predictable s.t. τ the process $\Delta C_t = \Delta M_\tau I_{\{\tau \le t\}} \in \mathcal{M}^2$, and for any $N \in \mathcal{M}^2$ the process*

$$L_t = C_t N_t - \Delta C_\tau \Delta N_\tau I_{\{\tau \le t\}} \in \mathcal{M}.$$

10.1 Basic elements of martingale theory

Proof The first claim follows from Remark 10.6. Further, for the process (L_t) we have

$$\sup_t |L_t| \leq \sup_t |C_t| \sup_t |N_t| + |\Delta C_\tau||\Delta N_\tau| \in L^1,$$

and for any s.t. σ and $N_{t \wedge \sigma} = N_t^\sigma$ by Lemma 10.3

$$\mathbf{E}C_\infty N_\infty^\sigma N = \mathbf{E}C_\sigma N_\sigma = \mathbf{E} \int_0^\infty N_s^\sigma dC_s = \mathbf{E}N_\tau^\sigma \Delta C_\tau$$
$$= \mathbf{E}\Delta N_\tau^\sigma \Delta C_\tau = \mathbf{E}\Delta N_\tau \Delta C_\tau I_{\{\sigma \leq \tau\}},$$

where the equality $\mathbf{E}N_{\sigma-}^\tau \Delta C_\sigma = \mathbf{E}\left[N_{\sigma-}^\tau \mathbf{E}(\Delta C_\sigma|\mathcal{F}_{\sigma-})\right] = 0$ was used. We arrive to conclusion that $\mathbf{E}L_\tau = 0$ for any s.t. τ. Now the claim follows from Lemma 10.4. □

Lemma 10.6 *Let $M \in \mathcal{M}^2$ and σ be a totally inaccessible s.t. Let $D_t = \Delta M_\sigma I_{\{\sigma\}}$ with compensator $\tilde{D}_t$ and $\hat{D}_t = D_t - \tilde{D}_t$. Then*
(1) The process $\hat{D} \in \mathcal{M}^2$ and $\mathbf{E}\hat{D}_\infty^2 \leq 5\mathbf{E}(\Delta M_\sigma)^2$;
(2) For any $N \in \mathcal{M}^2$ the process

$$L_t = D_t N_t - \Delta D_\sigma \Delta N_\sigma I_{\{\sigma \leq t\}} \in \mathcal{M}.$$

Proof (1) It is sufficient to consider $D_t = \phi I_{\{\sigma \leq t\}}$ with non-negative function ϕ, $\mathcal{F}_\sigma$-measurable and from L^2. Otherwise, one can consider separately ΔM_σ^+ and ΔM_σ^-. For function $|\phi| \leq a$ we have

$$\mathbf{E}\tilde{D}_\infty^2 = 2\mathbf{E}\int_0^\infty \tilde{D}_s d\tilde{D}_s = 2\mathbf{E}\int_0^t \tilde{D}_s dD_s$$
$$\leq 2\mathbf{E}\int_0^\infty \tilde{D}_\infty dD_s = 2\mathbf{E}\phi\tilde{D}_\infty \leq 2a\mathbf{E}\tilde{D}_\infty < \infty,$$

where the formula $f^2(\infty) = 2\int_0^\infty f(s)df(s)$ is used. Hence, $\tilde{D}_\infty \in L^2$ and $\mathbf{E}\tilde{D}_\infty^2 \leq 2\mathbf{E}\phi\tilde{D}_\infty \leq 2||\phi||_{L^2}||\tilde{D}_\infty||_{L^2}$ and $\mathbf{E}\tilde{D}_\infty^2 \leq 4\mathbf{E}\phi^2$ for a non-negative bounded function ϕ.

In general case, we consider $\phi^n = \phi \wedge n$ and processes $\tilde{D}_t^n$ and $D_t^n = \phi^n I_{\{\sigma \leq t\}}$. Correspondingly, $\tilde{D}_t^{n+1} - \tilde{D}_t^n$ is the compensator of the increasing continuous process we have

$$\mathbf{E}(\tilde{D}_\infty^n)^2 \leq 4\mathbf{E}(\phi^n)^2 \leq 4\mathbf{E}\phi^2 < \infty. \tag{10.14}$$

The non-decreasing sequence $\tilde{D}_\infty^n$ converges, and according to (10.14) the limit B_∞ will be in L^2, as in case $B_t = \lim_n \tilde{D}_t^n$. Each sample function $\tilde{D}_t^n(\omega)$ converges to $B_t(\omega)$ uniformly, as $0 \leq \tilde{D}_t^{n+} - \tilde{D}_t^n \leq \tilde{D}_\infty^{n+1} - \tilde{D}_\infty^n$. Hence, the process B_t is continuous, and by taking the limit in L^1 in the equality $\mathbf{E}(D_t^n - \tilde{D}_t^n|\mathcal{F}_s) = D_s^n - \tilde{D}_s^n$, $s \leq t$, we arrive to conclusion that $D_t - B_t \in \mathcal{M}$. Therefore, $B_t = \tilde{D}_t$, and $\mathbf{E}\tilde{D}_\infty^2 = \mathbf{E}B_\infty^2 \leq 4\mathbf{E}\phi^2 < \infty$ and the martingale $\hat{D} = D - \tilde{D} \in \mathcal{M}^2$.

(2) As in Lemma 10.5, the process L is uniformly integrable. Take a stopping time τ and applying Lemma 10.3 to $\hat{D}$ and N^τ, we obtain

$$\mathbf{E}(D_\tau - \tilde{D}_\tau)N_\tau = \mathbf{E}\left[\int_0^\infty N_t^\tau d(D-\tilde{D})_t\right] = \mathbf{E}N_\sigma^\tau \Delta D_\sigma - \mathbf{E}\int_0^\infty N_t^\tau d\tilde{D}_t$$

$$= \mathbf{E}N_\sigma^\tau \Delta D_\sigma - \mathbf{E}\left[\int_0^\infty N_{t-}^{\tau-} d\tilde{D}_t\right] = \mathbf{E}\left[\Delta N_\sigma \Delta D_\sigma I_{\{\sigma \le t\}}\right].$$

Finally, applying Lemma 10.4 we show that L is a martingale. □

Definition 10.11 A martingale $M \in \mathcal{M}^2$ is called *purely discontinuous*, if $M_0 = 0$, and if for any continuous martingale $N \in \mathcal{M}^2$ their product $M \cdot N$ is a martingale.

Let us denote $\mathcal{M}^{2,c}$ and $\mathcal{M}^{2,d}$ subclasses of $\mathcal{M}^2$ of continuous and purely discontinuous martingales. Now we are ready to formulate the key theorem for square integrable martingales.

Theorem 10.4 *Any martingale $M \in \mathcal{M}^2$ admits the following unique decomposition*

$$M = M^c + M^d, \tag{10.15}$$

where $M^c \in \mathcal{M}^{2,c}$ and $M^d \in \mathcal{M}^{2,d}$.

Let us call M^c in (10.15) the *continuous part of* M, and M^d the *compensated sum of the jumps of M* or the *purely discontinuous part of M*.

Proof Using Remark 10.6 with Lemma 10.5 and Lemma 10.6 we can see that $C^n \hat{D}^n$, $C^n \hat{D}^m$, $C^n C^m$, $\hat{D}^n \hat{D}^m$ ($m \ne m$) are martingales. Denote for $N = 1, 2, \ldots$, $Y_t^N = \sum_1^N C_t^n + \sum_1^N \hat{D}^n$ and find from Lemma 10.5 and Lemma 10.6 that for $N \le N'$:

$$\mathbf{E}(Y_\infty^{N'} - Y_\infty^N)^2 = \mathbf{E}\left[\sum_{N+1}^{N'}(C_\infty^n)^2 + \sum_{N+1}^{N'}(\hat{D}_\infty^n)^2\right] \le 5\mathbf{E}\left[\sum_{N+1}^{N'}\left((\Delta M_{\tau_n})^2 + (\Delta M_{\sigma_n})^2\right)\right]. \tag{10.16}$$

It follows from (10.16) and Theorem 9.4 that

$$\mathbf{E}\sup_t(Y_t^{N'} - Y_t^N)^2 \le 5\mathbf{E}\left[\sum_{N+1}^{N'}\left((\Delta M_{\tau_n})^2 + (\Delta M_{\sigma_n})^2\right)\right].$$

It implies $\sum_N \mathbf{E}\sup_t |Y_t^{N+1} - Y_t^N|^2 < \infty$ for some subsequence (Y^N), which we denoted Y^N again. By the Borel-Cantelli lemma, there exists $Y_t = \lim_N Y_t^N$ (a.s.) uniformly over t, and it is a limit in L^2 too. So, the process Y_t is cadlag with the same jumps as M, and $Y \in \mathcal{M}^2$. □

Now for a continuous martingale $X \in \mathcal{M}^2$ the process XY^N is a martingale by Lemma 10.5 and Lemma 10.6. The convergence Y_t^N to Y_t in L^2 implies a convergence $X_t Y_t^N$ to $X_t Y_t$ in L^1. Hence, it is a martingale, and $M^d = Y$ is a purely discontinuous martingale. Finally, $M^c = M - M^d$ is a continuous martingale, and we get (10.15).

To prove the *uniqueness* of such decomposition, we assume that there are two decompositions $M = M^c + M^d = \bar{M}^c + \bar{M}^d$, and $M^c - \bar{M}^c = \bar{M}^d - M^d$ is continuous and purely discontinuous. Hence, $(M_t^d - \bar{M}_t^d)^2 = \mathbf{E}(M_0^d - \bar{M}^d)^2 = 0$ for all $t \ge 0$.

Definition 10.12 Let $M \in \mathcal{M}^2$, $M_0 = 0$, M^c and M^d are its continuous and purely discontinuous parts of M. Define $[M, M]_t = \langle M^c, M^c \rangle_t + \sum_{s \leq t}(\Delta M_s)^2$ and call it a quadratic brackets of M.

We can note that $(M_t^c)^2 - \langle M^c, M^c \rangle_t$, $(M_t^d)^2 - \sum_{s \leq t}(\Delta M_s)^2$ and $M^c M^d$ belong to $\mathcal{M}$. Hence, $M^2 - [M, M] \in \mathcal{M}$, and $\langle M, M \rangle$ is the compensator of $[M, M]$.

Remark 10.7 If $M_0 \neq 0$, then we can think of M_0 as the jump of M at $t = 0$ and define
$$[M, M] = \langle M^c - M_0, M^c - M_0 \rangle + M_0^2 + \sum_{s \leq t}(\Delta M_s)^2.$$

Remark 10.8 Let us list some properties such quadratic characteristics of square integrable martingales.

Define for $M, N \in \mathcal{M}$, $M_0, N_0 = 0$, their *joint* quadratic characteristics (brackets, compensators):

$$\langle M, N \rangle_t = \frac{1}{2}(\langle M + N, M + N \rangle_t - \langle M, M \rangle_t - \langle N, N \rangle_t),$$

$$[M, N]_t = \frac{1}{2}([M + N, M + N]_t - [M, M]_t - [N, N]_t).$$

(1) The process $\langle M, N \rangle$ is the unique predictable process such that $MN - \langle M, N \rangle \in \mathcal{M}$.

(2) For any stopping time τ we have
$$\langle M^\tau, N \rangle = \langle M, N^\tau \rangle = \langle M^\tau, N^\tau \rangle = \langle M, N \rangle^\tau$$

and similar equalities are true for the quadratic brackets.

(3) $[M, N] = \langle M^c, N^c \rangle + \sum_{s \leq t} \Delta M_s \Delta N_s$.

(4) $MN - [M, N] \in \mathcal{M}$.

Analysing class $\mathcal{M}$ of square integrable martingales we observe that a Wiener process W does not belong to $\mathcal{M}$, because $\mathbf{E} W_t^2 = t \to \infty$ as $t \to \infty$. In the next section it will be shown how to extend $\mathcal{M}$ to include this key process to a bigger class for which a very nice theory can be developed.

10.2 Extension of martingale theory by localization of stochastic processes

We start with the following localization procedure. Let $\mathcal{H}$ be a class of stochastic processes $X = (X_t)_{t \geq 0}$. Define $\mathcal{H}_{loc}$ the set of stochastic processes $Y = (Y_t)_{t \geq 0}$ such that for each of them we can find a sequence of stopping times $\tau_n \uparrow \infty$, $n \to \infty$, such that the process $Y^{\tau_n} = (Y_{t \wedge \tau_n})_{t \geq 0}$ is in $\mathcal{H}$ for any $n = 1, 2, \ldots$. The sequence $(\tau_n)_{n=1,2,\ldots}$ is called *localizing (fundamental)* for the process Y. For example, in case of classes $\mathcal{M}_{loc}$ and $\mathcal{M}^2_{loc}$ we will call the corresponding processes *local martingales* and *locally square-integrable martingales*.

Definition 10.13 We say that a s.t. τ *reduces* $M \in \mathcal{M}_{loc}$ if $M^\tau \in \mathcal{M}$.

Problem 10.4 Prove that for $M \in \mathcal{M}_{loc}$ the following statements are true:
(1) A s.t. σ reduces $M \Leftrightarrow M^\sigma \in \mathcal{D}$.
(2) If a s.t. τ reduces M and a s.t. $\sigma \leq \tau$, then σ reduces M.
(3) If s.t.'s σ and τ reduce M, then $\max(\sigma, \tau) = \sigma \vee \tau$ reduces M.
Hint: to prove (3) one can use the equality

$$M^{\sigma \vee \tau} = M^\sigma + M^\tau - M^{\sigma \wedge \tau}.$$

The next theorem shows that the localization does not expand the class of local martingales.

Theorem 10.5 *If M is locally a local martingale, then M is a local martingale.*

Proof Denote a localizing sequence $(\tau_n)_{n=1,2,\ldots}$ such that M^{τ_n} is a local martingale. Denote $\mathcal{R}$ the set of all stopping times τ reducing M, and $\gamma = esssup_{\tau \in \mathcal{R}}\tau$. There exists a sequence $\sigma_n \in \mathcal{R}$ which converges to γ almost surely. Let us make the sequence $(\sigma_n)_{n=1,2,\ldots}$ non-decreasing with the help of (3) of Problem 10.4. By the definition γ majorates $\tau_n \to \infty$ $(n \to \infty)$, and, hence, $\gamma = \infty$ (a.s.), and $(\sigma)_{n \geq 1}$ is fundamental for M. □

Problem 10.5 (1) Let B be a process with locally integrable variation, i.e. $B \in \mathcal{A}_{loc}$. Prove that there exists a unique predictable process $A \in \mathcal{A}_{loc}$ such that $B - A \in \mathcal{M}_{loc}$. We call A a *compensator* of B.

(2) Prove that the predictable process B with finite variation has a *locally bounded variation*.

(3) A process with finite variation has a compensator $\Leftrightarrow$ its variation is locally integrable.

Remark 10.9 There is a simple way to check that the variation is locally integrable. Namely, if for the process A with finite variation there exists a localizing sequence (τ_n) such that $\mathbf{E}\sup_{s \leq \tau_n}|\Delta A_s| < \infty$, $n = 1, 2, \ldots$, then the variation of A is locally integrable. To verify this test we define a sequence of s.t.'s

$$\sigma_n = \tau_n \wedge \inf\left(t : \int_0^t |dA_s| \geq n\right), n = 1, 2, \ldots$$

and find that

$$\int_{[\![0,\sigma_n]\!]} |dA_s| \leq n + |\Delta A_{\sigma_n}| \leq n + \sup_{s \leq \tau_n} |\Delta A_s| \in L^1.$$

The next result plays a key role in many construction for local martingales. That is why it is called a *fundamental lemma* in the literature.

Lemma 10.7 *Let $M \in \mathcal{M}_{loc}$, then*
(1) $M_t^* = \sup_{s \leq t} |M_t| \in \mathcal{A}_{loc}^+$,
(2) $M_t = U_t + V_t$, where $U \in \mathcal{M}_{loc}, |\Delta U_t| \leq 1, t \geq 0, V \in \mathcal{M}_{loc} \cap \mathcal{V}$.

10.2 Extension of martingale theory by localization of stochastic processes

Proof (1) Take a localizing sequence of s.t.'s $(\sigma_n)_{n=1,2,...}$ for M, and make elements of this sequence finite as follows $\sigma_n \to \sigma_n \wedge n, n = 1, 2, \ldots$. Further, define s.t.'s

$$\tilde{\sigma}_n = \sigma_n \wedge \inf\{t : |M_t| \geq n\} \leq \sigma_n, \ n = 1, 2, \ldots,$$

and obtain that $M_{\tilde{\sigma}_n}$ is integrable due to the uniform integrability of $(M_{\sigma_n \wedge t})_{t \geq 0}$. On. stochastic interval $[\![0, \tilde{\sigma}_n [\![$ we have $|M_t| \leq n$ and $M^*_{\tilde{\sigma}_n} \leq n + |M_{\tilde{\sigma}_n}|$ belongs to L^1. Hence, $(M^*_t)_{t \geq 0}$ is locally integrable.

(2) Denote $(A_t) = \sum_{s \leq t} \Delta M_s I_{\{|\Delta M_s| > 1/2\}}$ which is finite for almost all ω. Using this process and $(\tilde{\sigma}_n)$, we define a new sequence of s.t.'s

$$\tau_n = \tilde{\sigma}_n \wedge \inf\{t : \int_0^t |dA_s| \geq n\} \leq \sigma_n, \ n = 1, 2, \ldots.$$

It is a localizing sequence such that

$$\int_{[\![0, \tau_n]\!]} |dA_t| \leq n + |\Delta A_{\tau_n}| \leq n + 2M^*_{\tau_n} \in L^1.$$

Therefore, there is a compensator B for A, and $V = A - B \in \mathcal{M}_{loc} \cap \mathcal{V}$. Because of predictability of B, its jumps are happen at predictable times τ only. Stopping processes by sequence $\tilde{\sigma}_n$ we get now

$$|\Delta B^{\tilde{\sigma}_n}_\tau| = \left|\mathbf{E}\left(\Delta A^{\tilde{\sigma}_n}_\tau | \mathcal{F}_{\tau-}\right)\right|$$
$$= \left|\mathbf{E}\left(\Delta M^{\tilde{\sigma}_n}_\tau - \Delta M^{\tilde{\sigma}_n}_\tau I_{\{|\Delta M_\tau| < 1/2\}} | \mathcal{F}_{\tau-}\right)\right| \leq 0 + 1/2.$$

Hence, the jumps of $U_t = M_t - V_t$ verify the following inequalities

$$|\Delta U_t| \leq |\Delta(M - A)_t| + |\Delta B_t| \leq 1.$$

Moreover, the sequence of stopping times defined as $\tilde{\tau}_n = \inf\{t : |U_t| \geq n\}$ is localizing, and $|U^{\tilde{\tau}_n}_t| \leq n + 1, n = 1, 2, \ldots$. □

Below we use this fundamental lemma to prove a decomposition of any local martingale into the sum of continuous and purely discontinuous if for any continuous local martingale $N \in \mathcal{M}_{loc}$ their product $MN \in \mathcal{M}_{loc}$.

Lemma 10.8 *Let $M \in \mathcal{M}_{loc} \cap \mathcal{V}$. Then*
(1) $V_t \sum_{s \leq t} \Delta M_s$ is a process with locally integrable variation and $M_t = V_t - \tilde{V}_t, t \geq 0$.
(2) For any bounded continuous martingale N, the product $MN \in \mathcal{M}_{loc}$.

Proof By Lemma 10.7 the process $M^*_t \in \mathcal{A}^+_{loc}$. By Remark 10.9 the variation of the process V is locally integrable. Therefore, there is a compensator $\tilde{V}$ for V, and the local martingale $M - (V - \tilde{V})$ is continuous, and its variation is locally integrable according to Lemma 10.7 and Remark 10.9. It means $M - (V - \tilde{V})$

admits a compensator, but due to a continuity of $M - (V - \tilde{V})$ it has to be zero as $M - (V - \tilde{V}) \in \mathcal{M}_{loc}$. As a result, $M = V - \tilde{V}$.

(2) Let N be a bounded continuous martingale and let the variations of V and $\tilde{V}$ are already integrable (otherwise, we can choose a localizing sequence of s.t.'s and work with stopped processes). For a s.t. τ with the help of Lemma 10.1 and Lemma 10.3 we obtain

$$\mathbf{E} M_\tau N_\tau = \mathbf{E} M_\infty N_\infty^\tau = \mathbf{E} \int_0^\infty N_s^\tau dM_s$$

$$= \mathbf{E} \int_0^\infty N_s^\tau dV_s - \mathbf{E} \int_0^\infty N_s^\tau d\tilde{V}_s$$

$$= \mathbf{E} \int_0^\infty N_s^\tau dV_s - \mathbf{E} \int_0^\infty N_s^\tau dV_s$$

$$= 0.$$

By Lemma 10.4 we arrive to conclusion that MN is a martingale. $\square$

Theorem 10.6 *Let $M \in \mathcal{M}_{loc}$, then*

$$M_t = M_t^c + M_t^d, \tag{10.17}$$

where $M^c \in \mathcal{M}_{loc}^c$ (continuous local martingales) and $M^d \in \mathcal{M}_{loc}^d$ (a class of purely discontinuous local martingales). The decomposition (10.17) *is unique.*

Proof According to the fundamental lemma we represent $M = N + U$ with a locally bounded local martingale N and a martingale with finite variation U. Taking a localizing sequence of s.t.'s (τ_n) we get N^{τ_n} as a bounded martingale for each $n = 1, 2, \ldots$. Applying Theorem 10.5 we obtain $N^{\tau_n} = (N^{\tau_n})^c + (N^{\tau_n})^d$, and using the uniqueness of such decomposition, we obtain that $(N^{\tau_n})^c = (N^{\tau_{n+1}})^c$ on $[\![0, \tau_n]\!]$. Now putting

$$N^c = \sum_{n=1}^\infty (N^{\tau_n})^c I_{]\!]\tau_{n-1}, \tau_n]\!]}$$

we get a continuous local martingale as well as a purely discontinuous local martingale $N^d = N - N^c$. As a result, we arrive to the following decomposition $M = N^c + (N^d + U)$, where $N^d + U$ is a purely discontinuous local martingale due to Lemma 10.8.

As far as the uniqueness of (10.17), we can assume that $M = X^c + X^d = M^c + M^d$. Then the equality $M^c - X^c = X^d - M^d$ leads to conclusion that locally $M^c - X^c = X^d - M^d = 0$ by Theorem 10.5. $\square$

Remark 10.10 For $M \in \mathcal{M} \in \mathcal{M}_{loc}^2$ we can apply the Doob-Meyer decomposition (for M^2) locally and construct its *compensator* $\langle M, M \rangle \in \mathcal{A}_{loc}^*$, which is predictable and $M^2 - \langle M, M \rangle \in \mathcal{M}_{loc}$. In particular such compensator is well defined for any continuous local martingale because $\mathcal{M}_{loc}^c \subseteq \mathcal{M}_{loc}^2$.

10.2 Extension of martingale theory by localization of stochastic processes 155

The exact extension of this approach to the whole class of local martingales is not possible. Nevertheless, one can define for $M \in \mathcal{M}_{loc}$ a similar *quadratic characterization* $[M, M]$ as follows

$$[M, M]_t = \langle M^c, M^c \rangle_t + \sum_{s \leq t} (\Delta M_s)^2,$$

where M^c is a continuous part of M.

The formula (10.2) shows that the process $[M, M]$ is increasing, $M^2 - [M, M] \in \mathcal{M}_{loc}$, but it is not predictable anymore.

To provide the correctness of the definition of process $[M, M]$ by formula (10.2) we need to prove that $\sum_{s \leq t} (\Delta M_s)^2$ *converges* for all t.

Again, we use the fundamental lemma to represent $M = U + V$. Further, $\sum_{s \leq t} |\Delta V_s|$ converges and therefore $\sum_{s \leq t} |\Delta V_s|^2$ converges too. Taking a localizing sequence (τ_n) for U we obtain that $\mathbf{E} \sum_{s \leq \tau_n} |\Delta U_s|^2 < \infty$ and for almost all ω

$$\sum_{s \leq \tau_n} |\Delta M_s| \leq 2 \sum_{s \leq \tau_n} |\Delta U_s|^2 + 2 \sum_{s \leq \tau_n} |\Delta V_s|^2$$

converges.

Let us join together classes of local martingales and processes with finite variation to create a bigger family of stochastic processes.

Definition 10.14 A process X is called a *semimartingale*, if

$$X = X_0 + M + A, \tag{10.18}$$

where X_0 is an $\mathcal{F}_0$-measurable r.v., $M \in \mathcal{M}_{loc}$, $A \in \mathcal{V}$.

The next auxiliary result speaks us when the semimartingale admits a unique representation in its definition.

Lemma 10.9 *Assume X is a semimartingale with bounded jumps $|\Delta X_t| \leq a < \infty$. Then X can be represented in a unique way as*

$$X = X_0 + M + A, \tag{10.19}$$

where X_0 is an $\mathcal{F}_0$-measurable r.v., $M \in \mathcal{M}_{loc}$, A is a predictable *process with finite variation.*

Proof Let us write the semimartingale $X = X_0 + N + B$ with $N \in \mathcal{M}_{loc}$ and $B \in \mathcal{V}$, and with the uniquely defined r.v. X_0-$\mathcal{F}_0$-measurable. We can estimate the jumps of B as follows

$$|\Delta B_t| \leq |\Delta N_t| + |\Delta X_t| \leq 2N_t^* + a, \ t \geq 0.$$

So, the increasing process $Y_t = \sup_{s \leq t} |\Delta B_s| \in \mathcal{A}_{loc}^+$, and by Remark 10.9 the variation of B is also locally integrable, and, hence, there exists its compensator A. Let us put $M = N + B - A \in \mathcal{M}_{loc}$ and find that $X = X_0 + M + A$. If there are two representations (10.19) for $X = X_0 + M + A = X_0 + M' + A'$, then $A - A' \in \mathcal{M}_{loc}$ and therefore A' is a compensator of A and due to predictability $A = A'$.

Now we are ready to prove that the class of semimartingales can not be extended by localization procedures.

Theorem 10.7 *Any stochastic process X which is locally a semimartingale is a semimartingale.*

Proof Denote $(\tau_n)_{n=1,2,...}$ a localizing sequence of stopping times, i.e. X^{τ_n} is a semimartingale for each $n = 1, 2, \ldots$. Define the process $Y_t = \sum_{s \leq t} \Delta X_s I_{\{|\Delta X_s| \geq 1\}}$ and note that Y has finite variation. Therefore, the difference $X^{\tau_n} - Y^{\tau_n}$ is a semimartingale with jumps bounded by one. Let us use Lemma 10.9 and write the unique representation $X^{\tau_n} - Y^{\tau_n} = X_0 + M^{(n)} + A^{(n)}$ with $M^{(n)} \in \mathcal{M}_{loc}$, $A^{(n)} \in \mathcal{V}$ and predictable. Further, using the uniqueness of such representation we obtain on $[\![0, \tau_n]\!]$ that

$$A^{(n)} = A^{(n+1)} \quad \text{and} \quad M^{(n)} = M^{(n+1)}.$$

So, we get the processes $A = \sum_n A^{(n)} I_{]\!]\tau_{n-1}, \tau_n]\!]}$ and $M = \sum_n M^{(n)} I_{]\!]\tau_{n-1}, \tau_n]\!]}$. $\square$

The process A is predictable process with finite variation, M is locally a local martingale, and according to Theorem 10.6 belongs to class $\mathcal{M}_{loc}$. As a result, we conclude that X is a semimartingale.

Another result of this type is presented in the following theorem.

Theorem 10.8 *Let X be a cadlag adapted process and there exist a sequence of s.t.'s (τ_n) and a sequence of semimartingales $Y^{(n)}$ such that*
(1) $\lim_{n \to \infty} \tau_n = \infty$ (a.s.)
(2) $X = Y^{(n)}$ on $[\![0, \tau_n[\![$, $n = 1, 2, \ldots$
Then X is a semimartingale.

Proof For each $n = 1, 2, \ldots$ we have

$$X_{t \wedge \tau_n} = Y^{(n)}_{t \wedge \tau_n} - Y^{(n)}_{\tau_n} I_{\{\tau_n \leq t\}} + X_{\tau_n} I_{\{\tau_N \leq t\}},$$

and therefore, X^{τ_n} is a semimartingale. Now, if we can transform (τ_n) to an increasing sequence, then we prove a semimartingale property of X according to Theorem 10.8. The trick to make it is standard. If τ and σ are s.t.'s, and X^σ and X^τ are semimartingales, then $X^{\sigma \wedge \tau}$ and $X^{\sigma \vee \tau} = X^\sigma + X^\tau - X^{\sigma \wedge \tau}$ are both semimartingales too. $\square$

Representation (10.18) for a semimartingale X may not be unique. Nevertheless, there is a continuous martingale component of X which does not depend on such representation. It gives a possibility to define a quadratic bracket of X that is a very helpful characteristic and tool for semimartingales. It is provided by the next theorem.

Theorem 10.9 *Let X be a semimartingale with representation (10.18).*
Then
(1) M^c in (10.18) does not depend on the particular representation (10.18);
(2) $\sum_{s \leq t}(\Delta X_s)^2$ converges a.s. for all $t \geq 0$.

10.2 Extension of martingale theory by localization of stochastic processes   157

Proof (1) Indeed, if $X = X_0 + M + A = X_0 + N + B$ with $M, N \in \mathcal{M}_{loc}$, $A, B \in \mathcal{V}$, then the difference $M - N \in \mathcal{M}_{loc} \cap \mathcal{V}$, and according to Lemma 10.8 $M - N$ is purely discontinuous and $(M - N)^c = 0$. Therefore, $M^c = N^c$ and we denote this component X^c and call it the *continuous local martingale part* of X. The proof of (2) is provided in the same way as for local martingales. □

Definition 10.15 Let X be a semimartingale. Then the increasing process

$$[X, X]_t = \langle X^c, X^c \rangle_t + \sum_{s \leq t} (\Delta X_s)^2, \quad t \geq 0, \tag{10.20}$$

is well defined and is called the *quadratic bracket* of X. If we have another semimartingale $Y = (Y_t)_{t \geq 0}$, we can define a joint bracket

$$[X, Y]_t = \langle X^c, Y^c \rangle_t + \sum_{s \leq t} \Delta X_s \Delta Y_s \tag{10.21}$$

with similar properties.

Example 10.2 Let X be a supermartingale, then it is a semimartingale. To show it we use Lemma 10.9. We consider the stepped supermartingale $X_t^n = X_{t \wedge n}$, $n = 1, 2, \ldots$ which can be represented as follows

$$X_t^n = \mathbf{E}(X_n | \mathcal{F}_t) + X_t^n - \mathbf{E}(X_n | \mathcal{F}_t) = \mathbf{E}(X_n | \mathcal{F}_t) + Z_t,$$

where Z is a non-negative supermartingale. For arbitrary s.t. σ we obtain $\mathbf{E}|Z_\sigma| < \infty$. Moreover, define s.t.'s

$$\tau_n = \inf(t : Z_t^* \geq n) \wedge n, \quad Z_t^* = \sup_{s \leq t} |Z_s|$$

and on $[\![0, \tau_n]\!]$ we have $Z_t^* \leq n + |Z_{\tau_n}| \in L^1$, which certifies that $Z \in D$, and by the Doob-Meyer decomposition (10.10) can be rewritten as $Z = M - A$ with a martingale M and a predictable process A with finite variation. Hence, Z is a semimartingale, and X is also a semimartingale.

Example 10.3 Assume X is a cadlag *process with independent increments*. Then $Z_t = \sum_{s \leq t} \Delta X_s I_{\{|\Delta X_s| \geq 1\}}$ is the process with independent increments and with finite variation. Define $Y = X - Z$ and observe that it has jumps bounded by one, and again, the process with independent increments. Hence, $Y_t - \mathbf{E}Y_t$ is a martingale, and we arrive to conclusion: X is a semimartingale if and only if $\mathbf{E}Y_t$ has the finite variation.

Example 10.4 Let X be an integrable right-continuous adapted process defined on $[0, \infty]$. We call it a *quasimartingale* on $[0, \infty]$, if

$$\mathbf{Var}(X) = \sup_{0 \leq t_0 < \ldots < t_n \leq \infty} \mathbf{E}\left[\sum_0^n \left|\mathbf{E}(X_{t_{i+1}} - X_{t_i} | \mathcal{F}_{t_i})\right|\right] < \infty.$$

The quasimartingale X admits the *Fisk decomposition* $X = M + Y - Z$, where M is a martingale, Y and Z are non-negative supermartingales on $[0, \infty]$. Hence, it is a semimartingale.

Now we want to discuss the question about semimartingales and change of equivalent probability measures.

Theorem 10.10 *If P and Q are equivalent probability measures on a measurable space $(\Omega, \mathcal{F})$, then each semimartingale with respect to P will be a semimartingale w.r. to Q.*

Proof is based on the following lemmas.

Lemma 10.10 *Let N be a positive local martingale w.r. to Q. Then N^{-1} is a semimartingale w.r. to Q.*

Proof If $N_t \geq a > 0$, then $N_t^{-1} \leq a^{-1}$, and therefore $\mathbf{E} N_t^{-1} < \infty$, $t \geq 0$. Note that x^{-1} is convex. By the Jensen inequality N_t^{-1} will be a submartingale, and hence, it is a semimartingale w.r. to Q by Example 10.2. In general case we consider a sequence of convex functions $f_n(x)$ such that $f_n(x) = \frac{1}{x}$ on $[1/n, \infty)$ and $[0, 1/n]$ the graph of f_n is the tangent to the curve $y = 1/x$ at the point $1/n$. For each $t \geq 0$ we have $\mathbf{E} f_n(N_t) < \infty$, which implies that $Y_n = f_n(N)$ is a submartingale w.r. to Q for each $n = 1, 2, \ldots$. Taking s.t.'s $\tau_n = \inf(t : N_t \leq 1/n)$ we get a localizing sequence such that $Y_n = 1/N$ on $[\![0, \tau_n]\!]$. So, by Theorem 10.9 $1/N$ is a semimartingale. □

Lemma 10.11 *The product of semimartingales is a semimartingales.*

Proof Let us prove it (it is enough) for X^2. We use $X = M + V$, where M is a locally bounded martingale, and V is a process with finite variation, as stated in the fundamental lemma.

Taking a localizing sequence $(\tau_n)_{n=1,2,\ldots}$ we get M^{τ_n} as a bounded martingale, V^{τ_n} with bounded variation on $[\![0, 1/n]\!]$. Let us show that $(M^{\tau_n} + B^{\tau_n})^2$ is a semimartingale. The term $(M^{\tau_n})^2$ is a submartingale. As far as $(V^{\tau_n})^2$ we consider a subdivision $0 \leq t_0 < t_1 < \ldots < t_n = t$ of $[0, t]$ and find that

$$\sum_i |(V_{t_{i+1}}^{\tau_n})^2 - (V_{t_i}^{\tau_n})^2| = \sum_i |V_{t_{i+1}}^{\tau_n} + V_{t_i}^{\tau_n}| |V_{t_{i+1}}^{\tau_n} - V_{t_i}^{\tau_n}|$$

$$\leq 2 \int_0^t |dV_s^{\tau_n}| \left(\sum_i |V_{t_{i+1}}^{\tau_n} - V_{t_i}^{\tau_n}| \right)$$

$$\leq 2 \left(\int_0^t |dV_s^{\tau_n}| \right)^2 .$$

To treat the term $M^{\tau_n} V^{\tau_n}$, we note that $M_t^{\tau_n} = \mathbf{E}(M_{\tau_n}^+ | \mathcal{F}_t) - \mathbf{E}(M_{\tau_n}^- | \mathcal{F}_t)$ and V^{τ_n} is the difference of two increasing processes. It means that we can consider only the case of non-negative M^{τ_n} and V^{τ_n}. The process $M_t^{\tau_n} \left(V_t^{\tau_n} - \Delta V_{\tau_n} I_{\{\tau_n \leq t\}} \right) = M_t^{\tau_n} U_t^n$ is a submartingale, and $M^{\tau_n} V^{\tau_n} = M^{\tau_n} U^n$ on $[\![0, \tau_n]\!]$. The application of Theorem 10.9 completes the proof of Lemma. □

Let us come back to the proof of Theorem. Denote $Z_\infty = \frac{dQ}{dP}$ on $(\Omega, \mathcal{F})$ and find that $Z_t = \mathbf{E}^P(Z_\infty|\mathcal{F}_t)$ is a martingale w.r. to P and Z_∞ is positive P and Q-a.s. Obviously, X is a local martingale w.r. to $P \Leftrightarrow XZ^{-1}$ is a local martingale w.r. to Q. Writing X as the product $\frac{X}{Z}Z$ we get the statement from Lemma 10.10 and Lemma 10.11.

Remark 10.11 We defined $[X, X]$ by (10.20). It turns out one can prove that $[X, X]_t = P - \lim_n \sum_1^n (X_{t_i} - X_{t_{i-1}})^2$, and it is invariant regarding equivalent changes of measures. Similar fact is true also for $[X, Y]$ (see (10.21)).

10.3 On stochastic calculus for semimartingales

We extend here Ito's calculus to bigger classes of processes defined on a standard stochastic basis $(\Omega, \mathcal{F}, (\mathcal{F}_t)_{t \geq 0}, P)$. A scheme of such extension will be presented sequentially for square-intgerable martingales, local martingales and semimartingales.

As we know (see, for example, Corollary 10.1 and Remark 10.7), any $M \in \mathcal{M}^2$ is described as $M_t = \mathbf{E}(M_\infty|\mathcal{F}_t)$, where $M_\infty \in L^2(\Omega, \mathcal{F}, P)$. It creates a possibility to think about $\mathcal{M}^2$ as a Hilbert space with the scalar product $(M, N) = \mathbf{E}M_\infty N_\infty$ and the norm $||M|| = (\mathbf{E}M_\infty^2)^{1/2}$ for $M, N \in \mathcal{M}^2$.

Let us fix $M \in \mathcal{M}^2$ and define

$$S^2(M) = \left\{ \phi : \phi \text{ is predictable}, \mathbf{E} \int_0^\infty \phi_s^2 d[M, M]_s < \infty \right\}$$

with the norm $||\phi||_{S^2(M)} = \left(\mathbf{E} \int_0^\infty \phi_s^2 d[M, M]_s \right)^{1/2}$.

In this space we consider a subspace $S^2_{step}(M)$ of predictable processes of the form

$$\phi = \phi_0^* I_{\{0\}} + \sum_{i=1}^{n-1} \phi_i I_{(t_i, t_{i+1}]},$$

where $\phi_0^* - \mathcal{F}_0$-measurable, $\phi_i - \mathcal{F}_{t_i}$-measurable, $i = 1, 2, \ldots, n$.

for $\phi \in S^2_{step}(M)$ we define the *stochastic integral of ϕ w.r. to M* as the process of $\mathcal{M}^2$:

$$(\phi \circ M)_t = \int_0^\infty I_{(s,t]}(s) \phi_s dM_s = \sum_{i=1}^n \phi_i (M_{t_{i+1} \wedge t} - M_{t_i \wedge t}), \ t \geq 0. \qquad (10.22)$$

Using relation (10.22) we can easily obtain the following *properties* of $\phi \circ M$:
(1) *Isometry*: $||\phi \circ M|| = ||\phi||_{S^2(M)}$.
(2) *Continuity* of $(\phi \circ M)_t$ if M is continuous.
(3) *Indistinguishability* of jumps $\Delta(\phi \circ M)_t$ and $\phi_t \Delta M_t$, $t > 0$.

The subspace $S^2_{step}(M)$ is dense in $S^2(M)$, and as in case of stochastic integration w.t. to a Wiener process, one can extend the definition of $\phi \circ M$ to the whole space $S^2(M)$ with the properties (1)-(3) above.

Now we want to provide a more tight connection of $\phi \circ M$ with quadratic brackets of M and $\phi \circ M$. Creating such a connection the following *Kunite-Watanabe inequality* is useful.

Lemma 10.12 *Let M and $N \in \mathcal{M}^2$, and ϕ and ψ be bounded measurable processes. Then*

$$\int_0^\infty |\phi_s||\psi_s||d[M,N]_s| \leq \left(\int_0^\infty \phi_s^2 d[M,M]_s\right)^{1/2} \left(\int_0^\infty \psi_s^2 d[N,N]_s\right)^{1/2}. \quad (10.23)$$

To prove (10.23), which is an analog of the Cauchy-Schwartz inequality, we consider for $\lambda \in \mathbb{R}$ and $s \leq t$ the difference

$$[M + \lambda N, M + \lambda N]_t - [M + \lambda N, M + \lambda N]_s.$$

It is non-negative, and, hence,

$$|[M,N]_t - [M,N]_s|^2 \leq ([M,M]_t - [M,M]_s)([N,N]_t - [N,N]_s).$$

Take $\phi = \sum_{i=1}^n \phi_i I_{(t_i, t_{i+1}]}$, $\psi = \sum_{j=1}^m \psi_j I_{(t_j, t_{j+1}]}$ with bounded r.v.'s ϕ_i and ψ_j, we obtain from the above inequality for brackets (10.23). After this we extend (10.23) to the class of all bounded measurable functions ϕ and ψ using limit arguments, as usual.

Theorem 10.11 *For $M \in \mathcal{M}^2$ and $\phi \in S^2(M)$ the stochastic integral $\phi \circ M$ is a unique $L \in \mathcal{M}^2$ such that for any $N \in \mathcal{M}^2$ and any predictable bounded ψ*
(1) $[L,N] = \phi \circ [M,N]$,
(2) $\psi \circ [\phi \circ M, N] = \psi\phi \circ [M,N] = [\psi\phi \circ M, N]$ and $\psi \circ (\phi \circ M) = \psi\phi \circ M$,
(3) $(\phi \circ M)^c = \phi \circ M^c$, $(\phi \circ M)^d = \phi \circ M^d$.
In particular, for a s.t. τ and process $\psi = I_{[\![0,\tau]\!]}$ we have equalities

$$(\phi \circ M)_\tau = \int_0^\tau \phi_s dM_s = (\psi \circ (\phi \circ M))_\infty = ((\psi\phi) \circ M)_\infty$$

$$= \int_0^\infty I_{[\![0,\tau]\!]} \phi_s dM_s.$$

Proof Due to the properties of $\phi \circ M$ and Lemma 10.12 the mapping $\phi \to \mathbf{E}\left((\phi \circ M)_\infty N_\infty - \int_0^\infty \phi_s d[M,N]_s\right)$ is continuous on $S^2(M)$ for any $N \in \mathcal{M}^2$.

Further, $I_t = (\phi \circ M)_t N_t - \int_0^t \phi_s d[M,N]_s$ is an uniformly integrable process, and for any s.t. τ $\mathbf{E}I_\tau = 0$. Hence, by Lemma 9.4 the process (I_t) is a martingale, and we arrive to the statement (1). As far as the uniqueness, assume L and $L' \in \mathcal{M}^2$ and $[L,N] = \phi \circ [M,N]$ for any $N \in \mathcal{M}$. Then we have $[L-L', N] = 0$ and

10.3 On stochastic calculus for semimartingales

$[L - L', L - L'] = 0$. Hence, $(L - L')^2$ is a martingale and we obtain equality $L = L'$. Other statements (2)-(3) are almost obvious. □

Now we show the way of construction of stochastic integral for $M \in \mathcal{M}_{loc}$ and a *locally bounded predictable process* ϕ, i.e. $\phi_{\tau_n} I_{(\tau_n > 0)}$ is bounded for a localizing sequence of s.t.'s $(\tau_n)_{n=1,2,\ldots}$.

To define such a stochastic integral $\phi \circ M$ we use the fundamental lemma, when $M = U + V$ with $U \in \mathcal{M}_{loc}^2$, V is a local martingale with bounded variation. Then for a localizing sequence $(\tau_n)_{n=1,2,\ldots}$ for ϕ, U, V we define

$$(\phi \circ M)_t = \int_0^t \phi_s^{\tau_n} dU_s^{\tau_n} + \int_0^t \phi_s^{\tau_n} dV_s^{\tau_n}$$

on $[\![0, \tau_n]\!]$. To continue the construction we need the following lemma.

Lemma 10.13 *(1) If V is a martingale with integrable variation and ϕ is predictable with* $\mathbf{E} \int_0^\infty |\phi_s| |dV_s| < \infty$, *then the Lebesgue-Stieltjes integral* $\int_0^t \phi_s dV_s \in \mathcal{M}$.

(2) if $V \in \mathcal{M}^2$ and ϕ is bounded, then the Stochastic integral $(\phi \circ V)_t$ and the Lebesgue-Stieltjes integral $\int_0^t \phi_s dV_s$ are indistinguishable.

Proof We prove statement (1) only because second statement can be proved in a similar manner. Define $S'(V)$ the set of predictable processes ϕ such that $\mathbf{E} \int_0^\infty |\phi_s| |dV_s| < \infty$. Then $\|\phi\|_{S'(V)} = \mathbf{E} \int_0^\infty |\phi_s| |dV_s|$ defines the norm of this space. If $\|\phi^n - \phi\|_{S'(V)} \to 0$, $n \to \infty$, then

$$\mathbf{E} \left\{ \sup_t \left| \int_0^t \phi_s^n dV_s - \int_0^t \phi_s dV_s \right| \right\} \leq \mathbf{E} \int_0^\infty |\phi_s^n - \phi_s| |dV_s|$$
$$\leq \|\phi^n - \phi\|_{S'(V)} \to 0, n \to \infty.$$

Using these findings we make a transition from step-predictable functions to all functions of space $S'(V)$. □

Coming back to the definition of $\phi \circ M$ by relation (10.3) one can concluded with the help of Lemma 10.13 that $\phi \circ M \in \mathcal{M}_{loc}$ and its definition does not depend on the decomposition $M = U + V$. Moreover, we arrive to statements which are similar to (1)-(3) of Theorem 10.12 if we replace class $\mathcal{M}^2$ by $\mathcal{M}_{loc}$.

The final step of construction of stochastic integral is the case of semimartingale. Assume ϕ is a predictable locally bounded process and X is a semimartingale of the form

$$X = X_0 + M + A, \tag{10.24}$$

where X_0 is a finite $\mathcal{F}_0$-measurable r.v., $M \in \mathcal{M}_{loc}$, $A \in \mathcal{V}$.

Let us define

$$(\phi \circ X)_t = (\phi \circ M)_t + \int_0^t \phi_s dA_s, \tag{10.25}$$

where $\int_0^t \phi_s dA_s$ is the Lebesgue-Stieltjes integral.

According to Lemma 10.13 the definition (10.25) of $(\phi \circ X)$ does not depend on the decomposition (10.24). As a result, let us formulate the list of natural properties of $\phi \circ X$:
(1) $\phi \circ X$ is a semimartingale,
(2) $(\phi \circ X)^c = \phi \circ X^c$,
(3) jumps $\Delta(\phi \circ X)_t$ and $\phi_t \Delta X_t$ are indistinguishable,
(4) $(\phi \circ X)_\tau = \int_0^\infty I_{[\![0,\tau]\!]} \phi_s dX_s = \int_0^\tau \phi_s dX_s$ for any finite s.t. τ.

Remark 10.12 It is interesting to note that there is no hope to go beyond semimartingales in extension of stochastic integration.

Let $X^1, \ldots, X^d$ be d semimartingales taking the values in $\mathbb{R}$. It is convenient to collect them to a d-dimensional process $X_t = (X_t^1, \ldots, X_t^d)$, called a d-valued semimartingale.

Let $F : \mathbb{B} = \mathbb{R}^d \to \mathbb{R}$, $d \geq 1$, be a twice continuously differentiable function. Then one can transform X with the help of this smooth function. It turns out the transformed process $F(X_t)$ will be a semimartingale satisfying the next *generalized Ito's formula*:

$$F(X_t) = F(X_0) + \sum_{i=1}^d \int_0^t D^i F(X_{s-}) dX_s^{i,c}$$
$$+ \frac{1}{2} \sum_{1 \leq i,j, \leq d} \int_0^t D^i D^j F(X_{s-}) d[X^{i,c}, X^{j,c}]_s \qquad (10.26)$$
$$+ \sum_{s \leq t} \left(F(X_s) - F(X_{s-}) - \sum_{i=1}^d D^i F(X_{s-}) \Delta X_s^i \right),$$

where $D^i F(x) = \frac{\partial F(x)}{\partial_i} = \frac{\partial F(x^1, \ldots, x^d)}{\partial x_i}$, $D^i D^j F(x) = \frac{\partial^2 F(x)}{\partial x^i \partial x^j}$, $i, j = 1, \ldots, d$.

Remark 10.13 Let us note that processes $D^i F(X_{s-})$ and $D^i D^j F(X_{s-})$ are predictable and locally bounded, and therefore all integrals in (10.26) are well-defined.

Defining the localizing sequence $\tau_n = \inf\{t : |X_t| \geq n\}$, we obtain that $|X_t|$ and $|X_{t-}| \leq n$ on $[\![0, \tau_n[\![$, $n = 1, 2, \ldots$ Further, letting $K_n = \sup_{|x| \leq n} \sum_{i,j} |D^i D^j F(x)|$, weget with the help of the Taylor decomposition that

$$|\sum_{s < \tau_n} (F(X_s) - F(X_{s-}) - \sum_{i=1}^d D^i F(X_{s-}) \Delta X_s^i)| \leq \frac{K_n}{2} \sum_i \sum_{s \tau_n} |\Delta X_s^i|^2.$$

Hence, for almost all ω both series

$$\sum_{s < \tau_n} |F(X_s) - F(X_{s-}) - \sum_{i=1}^d D^i F(X_{s-}) \Delta X_s^i|$$

and

10.3 On stochastic calculus for semimartingales

$$\sum_{s \leq \tau_n} |F(X_s) - F(X_{s-}) - \sum_{i=1}^{d} D^i F(X_{s-}) \Delta X_s^i|$$

converge, and we can conclude that

$$\sum_{s \leq t} |F(X_s) - F(X_{s-}) - \sum_{i=1}^{d} D^i F(X_{s-}) \Delta X_s^i|$$

converges for all $t > 0$.

This observation generates the idea that the formula (10.26) can be proved without big technical difficulties in between moments of jumps, i.e. in a *pure continuous case*.

This is the reason that we prove (10.26) for a one-dimensional continuous semimartingale $X_t = X_0 + M_t + A_t$, $M \in \mathcal{M}^c_{loc}$, $A \in \mathcal{V}$. Using a localization procedure we can reduce the proof to the case when continuous processes $M, A, [M, M]$ are bounded as well as X_0, F, F' and F'', and the upper bound for $|F|, |F'|, |F''|$ is denoted by C.

Let us use the Taylor decomposition

$$F(y) - F(x) = (y - x)F'(x) + \frac{1}{2}(y - x)^2 F''(x) + o(x, y)$$

where $o(x, y) \leq \epsilon(|y - x|)|y - x|^2$ and ϵ is a non-decreasing function with $\lim_{t \to 0} \epsilon(t) = 0$.

Take $a > 0$ and define a sequence of s.t.'s $(\tau_n)_{n=0,1,\ldots}$ as follows

$$\tau_0 = 0, \ldots, \tau_{i+1} = t \wedge (\tau_i + a) \wedge \inf\{s > \tau_i : M_s - M_{\tau_i} > a,$$
$$[M, M]_s - [M, M]_{\tau_i} > a \text{ or } |A_s - A_{\tau_i}| > a\}, \ldots$$

Obviously, for each $\omega : \tau_i(\omega) = t$ except for a finite number of i, and we can write the following equality using the Taylor formula above:

$$F(X_t) - F(X_0) = \sum_i \left(F(X_{\tau_{i+1}}) - F(X_{\tau_i}) \right)$$
$$= \sum_i \left[F'(X_{\tau_i})(X_{\tau_{i+1}} - X_{\tau_i}) + \frac{1}{2} F''(X_{\tau_{i+1}})(X_{\tau_{i+1}} - X_{\tau_i})^2 + o(X_{\tau_i}, X_{\tau_{i+1}}) \right].$$
(10.27)

For the martingale part of (10.27) we have

$$\mathbf{E}\left[\int_0^t F'(X_s)dM_s - \sum_i F'(X_{\tau_i})(M_{\tau_{i+1}} - M_{\tau_i})\right]^2$$
$$=\mathbf{E}\left[\sum_i \int_{\tau_i}^{\tau_{i+1}} (F'(X_s) - F'(X_{\tau_i}))^2 \, d[M,M]_s\right] \qquad (10.28)$$
$$\leq \mathbf{E}\left(\sup_i |F'(X_s) - F'(X_{\tau_i})|^2 \cdot [M,M]_t\right) \to 0$$

as $a \to 0$.

The term of (10.27) with process A can be treated similarly, and $\sum_i F'(X_{\tau_i})(A_{\tau_i} - A_{\tau_i})$ converges to $\int_0^t F'(X_s)dA_s$ in L^1-sense.

The second term in (10.27) we represent as a sum of three components.

For the first component we have

$$|\sum_i F''(X_{\tau_i})(A_{\tau_{i+1}} - A_{\tau_i})^2| \leq C \sup_i |A_{\tau_{i+1}} - A_{\tau_i}| \int_0^t |dA_s|$$
$$\leq Ca \int_0^t |dA_s| \to 0, \; a \to . \qquad (10.29)$$

for the second component we obtain

$$|\sum_i F''(X_{\tau_i})(A_{\tau_{i+1}}) - A_{\tau_i})(M_{\tau_{i+1}} - M_{\tau_i})| \leq C \sup_i |M_{\tau_{i+1}} - M_{\tau_i}| \int_0^t |dA_s|$$
$$\leq Ca \int_0^t |dA_s| \to 0, \; a \to 0. \qquad (10.30)$$

Taking into account that $M^2 - [M,M]$ is a martingale, we get for the third component that

$$\mathbf{E}\left[\left(\sum_i F''(X_{\tau_i})(M_{\tau_{i+1}} - M_{\tau_i})^2 - \sum_i F''(X_{\tau_i})([M,M]_{\tau_{i+1}} - [M,M]_{\tau_i})\right)^2\right]$$
$$= \sum_i \mathbf{E}\left[(F''(X_{\tau_i}))^2 \left((M_{\tau_{i+1}} - M_{\tau_i})^2 - ([M,M]_{\tau_{i+1}} - [M,M]_{\tau_i})\right)^2\right]$$
$$\leq 2C^2 \mathbf{E}\left\{\sup_i |M_{\tau_{i+1}} - M_{\tau_i}|^2 M_t^2\right\} + 2C^2 \mathbf{E}\left\{\sup_i |[M,M]_{\tau_{i+1}} - [M,M]_{\tau_i}| \cdot [M,M]_t\right\}$$
$$\leq 2C^2 a^2 \mathbf{E} M_t^2 + 2C^2 a \mathbf{E}[M,M]_t \to 0, \; a \to 0. \qquad (10.31)$$

The remaining term in (10.27) can be treated as follows

10.3 On stochastic calculus for semimartingales

$$\mathbf{E}\sum_i o(X_{\tau_i}, X_{\tau_{i+1}}) \leq \mathbf{E}\left[\sum_i (X_{\tau_{i+1}} - X_{\tau_i})^2 \epsilon(|X_{\tau_{i+1}} - X_{\tau_i}|)\right]$$

$$\leq \mathbf{E}\left[2\epsilon(2a)\sum_i (A_{\tau_{i+1}} - A_{\tau_i})^2 + 2\epsilon(2a)\sum_i (M_{\tau_{i+1}} - M_{\tau_i})^2\right]$$

$$\leq 2\epsilon(2a)\mathbf{E}\left\{a\int_0^t |dA_s| + M_t^2\right\} \to 0, \ a \to 0. \tag{10.32}$$

Putting together relations (10.28)-(10.32) we complete the proof.

Now we are ready to consider stochastic differential equations with respect to semimartingale. The main result is contained in the next theorem.

Theorem 10.12 *Let $N = M + A$ be a semimartingale, $M \in \mathcal{M}_{loc}$, $A \in \mathcal{V}$. Assume function $F : \mathbb{R}_+ \times \Omega \times \mathbb{R} \to \mathbb{R}$ satisfies the conditions*
  (1) $f(s, \omega, \cdot)$ is a Lipschitz function with a constant K;
  (2) $f(s, \cdot, x)$ is $\mathcal{F}_s$-measurable;
  (3) $f(\cdot, \omega, \cdot)$ is a continuous function.
Then the SDE w.r. to N

$$X_t(\omega) = X_0(\omega) + \int_0^t f(s, \omega, X_{s-}(\omega))dN_s(\omega) \tag{10.33}$$

has a unique solution as a cadlag adapted process with the initial value $X_0(\omega) - \mathcal{F}_0$-measurable finite random variable.

Proof Obviously, (1)-(3) guarantee that the integral in the right-hand side of (10.33) is well-defined. The proof includes a couple of steps. We start with the assumption that processes

$$[M, M]_t \text{ and } \int_0^t |dA_s| \text{ are bounded by } b > 0 \tag{10.34}$$

and

$$|f(t, \omega, 0)| \leq c \text{ for all } t, \omega. \tag{10.35}$$

Let H be a class of cadlag processes X such that $X^* = \sup_t |X_t| \in L^2$, $X_0 = 0$. Define the norm in this space $||X|| = ||X^*||_{L^2}$.

Denote for $x \in H$

$$U(X)_t = \int_0^t f(s, \omega, X_{s-})dN_s.$$

Lemma 10.14 *The process $U(x) \in H$ if $X \in H$. Moreover, for processes X and $Y \in H$ we have*

$$||U(X) - U(Y)|| \leq h||X - Y||, \ h = K(2\sqrt{b} + b). \tag{10.36}$$

Proof For zero-process we have
$$U(0)_t = \int_0^t f(s,\omega,0)dM_s + \int_0^t f(s,\omega,0)dA_s = L_t + V_t.$$

Process $L \in \mathcal{M}_{loc}$, and due to (10.34)-(10.35)
$$[L,L]_\infty = \int_0^\infty f^2(s,\omega,0)d[M,M]_s \leq c^2 b, \qquad (10.37)$$

and $\mathbf{E}(L^*)^2 \leq 4c^2 b$ by Doob's inequality. □

For process V_t we get from (10.34)-(10.35) that
$$V^* \leq \int_0^\infty |dV_s| \leq \int_0^\infty |f(s,\omega,0)||dA_s| \leq cb \qquad (10.38)$$

and hence $\mathbf{E}(V^*)^2 \leq c^2 b^2$. Relations (10.37)-(10.38) show that $U(X) \in H$. Further, denote $Z = X - Y$ and find that

$$U(X)_t - U(Y)_t = \int_0^t (f(X_{s-}) - f(Y_{s-}))dM_s + \int_0^t (f(X_{s-}) - f(Y_{s-}))dA_s$$
$$= L'_t + V'_t, \; L' \in \mathcal{M}_{loc}, \; V' \in \mathcal{V}.$$

Using similar reasonings as in (10.37)-(10.38) we obtain for processes L' and V' that □

$$[L',L']_\infty = \int_0^\infty (f(X_{s-}) - f(Y_{s-}))^2 d[M,M]_s \leq K^2 b(Z^*)^2, \qquad (10.39)$$

$$\mathbf{E}((L')^*)^2 \leq 4K^2 b \mathbf{E}(Z^*)^2 \text{ and}$$

$$V^* \leq \int_0^\infty K|Z_s||dA_s| \leq KbZ^*. \qquad (10.40)$$

It follows from (10.39)-(10.40) that
$$\|U(X) - U(Y)\| \leq K\|X-Y\|(b + 2\sqrt{b}),$$

and we get (10.36).

Lemma 10.15 *Assume conditions* (10.34)-(10.35) *are satisfied, and* $h = K(b + 2\sqrt{b}) < 1$. *Then there exists a unique cadlag adapted process X solving the equation*

$$X_t = \int_0^t f(s,\omega, X_{s-})dN_s.$$

Proof Due to $h < 1$ there exists a unique solution X from space H. Let us note from conditions (1)-(2) of the theorem that
$$|\Delta N_s| \leq |\Delta M_s| + |\Delta A_s| = (\Delta[M,M]_s)^{1/2} + |\Delta A_s| \leq \sqrt{2b} + b.$$

10.3 On stochastic calculus for semimartingales

So, if Z is a cadlag adapted process, $Z_t = \int_0^t f(s, \omega, Z_{s-})dN_s$, we can define stopping times $\tau_n^Z = \inf(t : |Z_t| \geq n), n = 1, 2, \ldots$, and get

$$\Delta Z_{\tau_n^Z} = f(\tau_n^Z, Z_{\tau_n^Z-})\Delta N_{\tau_n^Z} \text{ and}$$

$$|\Delta Z_{\tau_n^Z}| \leq (c + nK)(2b + \sqrt{2b}).$$

Hence, the process Z is locally bounded and locally in space H. Due to uniqueness in H we obtain $Z = X$. □

Lemma 10.16 *Assume that condition (10.34) is fulfilled and $h = K(b + 2\sqrt{b}) < 1$, then there exists a unique cadlag adapted process X solving equation*

$$X_t = \int_0^t f(s, \omega, X_{s-})dN_s.$$

Proof Let $\tau_n = \inf(t : |f(t, \omega, 0)| \geq n)$, $n = 1, 2, \ldots$ and $f_n(t, \omega, x) = f(t, \omega, x)I_{\{0 < t \leq \tau_n\}}$. Obviously, functions f_n satisfy (10.35) with $c = n$, and each equation $Y_t^n = \int_0^t f_n(s, \omega, Y_{s-}^n)dN_s$ has a unique solution. Due to $f_{n+1} = f_n$ on $[\![0, \tau_n]\!]$ we have $Y^n = Y^{n+1}$ on $[\![0, \tau_n]\!]$.

Further, an adapted process X solves the equation $X_t = \int_0^t f(s, \omega, X_{s-})dN_s \Leftrightarrow$ for each τ_n we have $X_t^{\tau_n} = \int_0^t f(s, \omega, X_{s-}^{\tau_n})N_s^{\tau_n}$. Let us define now the process X as follows $X = Y^n$ on $[\![0, \tau_n]\!]$, $X_0 = 0$, and find the unique solution. □

Lemma 10.17 *Assume a semimartingale N has jumps $|\Delta N_t| \leq b/4$ and both b, $h = K(b + 2\sqrt{b}) < 1$. Then the equation $X_t = \int_0^t f(s, \omega, X_{s-})dN_s$ has one and only one solution.*

Proof Under the conditions of Lemma we can write $N = M + A$, $M \in \mathcal{M}_{loc}$ and A is a predictable process with locally bounded variation. Without loss of generality we can assume that $M \in \mathcal{M}$ and predictable process A has integrable variation. Then

$$|\Delta A_\tau| = |\mathbf{E}(\Delta N_\tau|\mathcal{F}_{\tau-}) - \mathbf{E}(\Delta M_\tau|\mathcal{F}_{\tau-})| = |\mathbf{E}\Delta N_\tau|\mathcal{F}_{\tau-})| \leq b/4$$

and we have a boundedness of $\Delta[M, M]_t = (\Delta M_T)^2$ and $|\Delta A_t|$ by $b/4$. Denote $D_t = \int_0^t |dA_s|$ and put $\tau_0 = 0, \ldots, \tau_n = \inf(t > \tau_{n-1} : D_t - D_{\tau_{n-1}} \geq b/2)$. We have $\int_{]\!]\tau_n, \tau_{n+1}]\!]} d[M, M]_s \leq b$ and $\int_{]\!]\tau_n, \tau_{n+1}]\!]} |dA_s| \leq b$ for all $n = 1, 2, \ldots$

By Lemma 9.16 there is one and only one solution on $[\![0, \tau_1]\!]$, and denote it by X^1. At the interval $]\!]\tau_1, \tau_2]\!]$ we have equation

$$X_t = X_{\tau_1}^1 + \int_{]\!]\tau_1, t]\!]} f(s, \omega, X_{s-})dN_s.$$

Obviously, X solves (86) $\Leftrightarrow Y = X - X_{\tau_1}^1$ is a solution of the equation

$$Y_t = \int_0^t f(s, \omega, Y_{s-} + X_{\tau_1}^1)dN_s.$$

So, we get a unique solution X^2 at this interval and so on. □

Let us come back to the proof of the theorem. We assume that b as in Lemma 10.17 and we have a non-decreasing sequence of s.t.'s $\tau_1 \leq \tau_2 \leq \ldots \leq \tau_n \leq \ldots$ at which $\Delta N_t| > b/4$. Construct the following semimartingale

$$N_t^1 = M_t I_{(t<\tau_1)} + N_{\tau_1-} I_{(t\geq\tau_1)}.$$

A cadlag process X is a solution of (10.33) on $[\![0, \tau_1]\!] \Leftrightarrow$ the process $X_t^1 = X_t I_{(t<\tau_1)} + X_{\tau_1-} I_{(t\geq\tau_1)}$ is a solution on $[\![0, \tau_1]\!]$ of the equation $X_t^1 = X_0 \int_0^t f(s, \omega, X_{s-}^1) dN_s^1$. On the interval $[\![0, \tau_1]\!]$ the process N^1 satisfies the conditions of Lemma 10.17, and hence the last SDE has a unique cadlag adapted solution X^1 on $[\![0, \tau_1]\!]$. We can continue this construction on $[\![\tau_1, \tau_2]\!]$ and so on.

Example 10.5 Let us consider the linear SDE:

$$Z_t = 1 + \int_0^t Z_{s-} dN_s. \tag{10.41}$$

We know from Theorem 10.12 that a unique solution of the equation (10.41) exists. It turns out it admits a special form which is similar to the Girsanov exponent:

$$Z_t = \exp\{N_t - \frac{1}{2}[N^c, N^c]_t\} \prod_{s \leq t} (1 + \Delta N_s) e^{-\Delta N_s}. \tag{10.42}$$

The form of solution in (10.42) is called a *stochastic exponent* or the Doleans exponent, denoted by $\mathcal{E}_t(N)$. To prove this, we denote $Y_t = N_t - \frac{1}{2}[N^c, N^c]_t$ and $X_t = \prod_{s \leq t}(1 + \Delta N_s)e^{-\Delta N_s}$, and apply Ito's formula to the function $F(Y_t, X_t) = \exp(Y_t)X_t$.

Problem 10.6 Prove the properties of stochastic exponents w.r. to semimartingales:
(1) Let $\mathcal{E}_t(N)$ be a stochastic exponent w.r. to a semimartingale N with $\Delta N_t \neq -1$, $t \geq 0$. Then $\frac{1}{\mathcal{E}_t(N)} = \mathcal{E}_t(-N^*)$, where $N_t^* = N_t - [N^c, N^c]_t - \sum_{s \leq t} \frac{(\Delta N_s)^2}{1+\Delta N_s}$. stochastic exponent w.r. to a semimartingale N with $\Delta N_t \neq -1$, $t \geq 0$. Then $\frac{1}{\mathcal{E}_t(N)} = \mathcal{E}_t(-N^*)$, where $N_t^* = N_t - [N^c, N^c]_t - \sum_{s \leq t} \frac{(\Delta N_s)^2}{1+\Delta N_s}$.
(2) If N in $\mathcal{M}_{loc}$, then $\mathcal{E}_t(N) \in \mathcal{M}_{loc}$.
(3) $\mathcal{E}_t(N) = 0$ for $(t, \omega) \in [\![\tau, \infty[\![$, where $\tau = \inf(t : \Delta N_t = -1)$.
(4) $\mathcal{E}_t(U)\mathcal{E}_t(V) = \mathcal{E}_t(U + V + [U, V])$, where U and V are semimartingale. This property is called the *rule of multiplication of stochastic exponents*.

Remark 10.14 The equation (10.33) admits further generalizations. One can prove an analog of Theorem 9.13 for equation

$$dX_t(\omega) = b(t, \omega, X_{t-}) dA_t + \sigma(t, \omega, X_{t-}) dM_t, \tag{10.43}$$

where $M \in \mathcal{M}_{loc}$, $A \in \mathcal{V}$, both coefficients b and σ are Lipschitz's functions, and $b(t, \omega, x)$ is optional and $\sigma(t, \omega, x)$ is predictable as functions of (t, ω).

Let us note that the equation 10.43 can be also considered for a multidimensional semimartingale $X = A + M$, and for matrix-valued coefficients b and σ.

Remark 10.15 The equation (10.33) admits further important results. One can prove an analog of comparison Lemma 9.1 for stochastic differential equations (10.43) with a predictable increasing process $A = (A_t)_{t \geq 0}$.
Moreover the Ito's formula (10.26) admits a further generalization to a class of convex functions F. The corresponding formula is similar to (10.26) where the second term in the right hand side is replaced by a continuous process V from $\mathcal{V}$. The process V admits a representation with the help of a *local time of X*.

10.4 The Doob-Meyer decomposition: proof and related remarks

The leading idea of this short book as it was stated in preface dictates the way of exposition avoiding too long proofs and too many technical details. That is why the proof of the Doob-Meyer decomposition was omitted in Section 10.1. However, the proof of this fundamental fact of stochastic analysis must be presented in an appropriate manner.
It is necessary to note that the initial Doob-Meyer decomposition in continuous time was given by Meyer in the form

$$X_t = X_0 + M_t + A_t, \quad t \geq 0, \tag{10.44}$$

where $X = (X_t)_{t \geq 0}$ is a submartingale of class (D), $M = (M_t)_{t \geq 0}$, $M_0 = 0$, is a martingale, $A = (A_t)_{t \geq 0}$, $A_0 = 0$, is a "natural" increasing process.
The notion "natural" means that for every bounded cadlag martingale $m = (m_t)_{m \geq 0}$ on a standard stochastic basis $(\Omega, F, (F_t)_{t \geq 0}, P)$ on the following equality is satisfied

$$\mathbf{E} \int_0^t m_s \, dA_s = \mathbf{E} \int_0^t m_{s-} \, dA_s. \tag{10.45}$$

This characterization of A looks very artificial in comparison with discrete time, where A is predictable. Nevertheless, since that time there were done several proofs of the Doob-Meyer decomposition in these terms, including the proof of Rao who used the limiting arguments for transition from discrete time to continuous time decomposition. Doleans was the first who noted that the notions "natural" and "predictable" are equivalent in (10.44).
This section is devoted to the proof of (10.44) in such a new fashion, and as in many other books we work with this form only. Nevertheless, before providing the proof which is based on the paper [2], we want to compare notions "natural" and "predictable".
First of all, one can show that $\mathbf{E} m_t A_t = \mathbf{E} \int_0^t m_s \, dA_s$ and therefore, the condition (10.45) is equivalent to the next one

$$Em_t A_t = \mathbf{E}\int_0^t m_{s-} dA_s. \tag{10.46}$$

Problem 10.7 Prove (for discrete time) that an increasing process A is natural $\Longleftrightarrow$ A is predictable, i.e. A_n-F_{n-1}-measurable, $n \geq 1$.

Hint: We rewrite (10.46) for the discrete time case as follows
$$Em_n A_n = \mathbf{E}\sum_1^n m_{k-1}\Delta A_k.$$

It follows from this condition that $Em_n(A_n - \mathbf{E}(A_n|F_{n-1})) = 0$ and even $\mathbf{E}|A_n - \mathbf{E}(A_n|F_{n-1})| = 0$ which leads to predictability of A.

The opposite implication follows from the equality $\mathbf{E}\sum_1^n A_k \Delta m_k = 0$ and the representation
$$\sum_1^n A_k \Delta m_k = A_n m_n - m_{n-1}\Delta A_n - m_{n-2}\Delta A_{n-1} - \ldots$$

which leads to $\mathbf{E}A_n m_n = \mathbf{E}\sum_1^n m_{k-1}\Delta A_k$.

Let us show that the decomposition (10.44) is unique in the class of natural processes A. To prove it we assume that there exist two such decompositions for X with martingales M', M'', and with the natural increasing processes A', and A''. Hence, $M'_t + A'_t = M''_t + A''_t$ for all $t \geq 0$. Define process $N_t = A'_t - A''_t = M'_t - M''_t$, which is a martingale with finite variation.

Further, for every bounded martingale $m = (m_t)_{t\geq 0}$, we have due to (10.46) that
$$Em_t.(A'_t - A''_t) = \mathbf{E}\int_0^t m_{s-}dN_s = \lim_{n\to\infty}\mathbf{E}\sum_j^{k_n} m_{t_{j-1}^{(n)}}(N_{t_j^{(n)}} - N_{t_{j-1}^{(n)}}),$$

where $(t_k^{(n)})_{k=0,\ldots,k_n}$ is a sequence of subdivision of $(0,T]$ with diameter $\Delta t_k^{(n)} \to 0$, $n \to \infty$. Obviously, $Em_{t_{j-1}^{(n)}}\Delta N_{t_j^{(n)}} = 0$, and, hence, $Em_t.(A'_t - A''_t) = 0$, $t \geq 0$. To finish the proof we take an arbitrary integrable random variable ζ and define a martingale $\zeta_t = \mathbf{E}(\zeta|F_t)$. Then $\mathbf{E}\zeta.(A'_t - A''_t) = 0$ for every such random variable which leads to conclusion that $A'_t = A''_t$ (a.s) for all t.

Let us prove the Doob-Meyer decomposition (10.44) assuming for simplicity that $t \in [0, 1]$.

Denote $D = \{D_1, D_2, \ldots, D_n, \ldots\}$, where $n = 1, 2, \ldots, D_n = \{0, \frac{1}{2^n}, \frac{2}{2^n}, \ldots, \frac{j}{2^j}, \ldots\}$, the dyadic numbers of the interval $[0, 1]$.

Consider the process $X = (X_t)_{t\geq 0}$ on this discrete time, and write the Doob-Meyer decomposition for $t \in D_n$:

$$X_t = M_t^n + A_t^n, \tag{10.47}$$

$$A_t^n - A_{t-\frac{1}{2^n}}^n = \mathbf{E}(X_t - X_{t-\frac{1}{2^n}}|F_{t-\frac{1}{2^n}}),$$

$$M_t^n = X_t - A_t^n.$$

10.4 The Doob-Meyer decomposition: proof and related remarks

Here $M^n = (M_t^n)_{t \in D_n}$ is a martingale, $A^n = (A_t^n)_{t \in D_n}$ is increasing and predictable with respect to $(F_t)_{t \in D_n}$.
The idea is to take a limit in (10.47), and as a result, to obtain (10.44).
To make such limiting transition carefully we need the following lemma.

Lemma 10.18 *Assume $(f_n)_{n \geq 1}$ is a uniformly integrable sequence of random variables on the probability space (Ω, F, P). Then there exists $g_n \in conv\{f_n, f_{n+1}, ...\}$ which converges in space $L^1(\Omega, F, P)$.*

We will use here a version of the Komlos Lemma:
If $(f_n)_{n \geq 1}$ is a bounded sequence on a Hilbert space, then for each n one can pick $g_n \in conv\{f_n, f_{n+1}, ...\}$ such that $(g_n)_{n \geq 1}$ will converge in the norm $\|.\|_2$ of this space.
To Prove this lemma, we define the finite number

$$a = \sup_n \inf\{\|g\|_2 : g \in conv\{f_n, f_{n+1}, ...\}\}$$

For each n we can take some $g_n \in conv\{f_n, f_{n+1}, ...\}$ with the norm $\|g_n\|_2 \leq a + \frac{1}{n}$.
For a fixed $\varepsilon > 0$, and a big enough n we obtain for all $m, k \geq n$ that

$$\|(g_k + g_m)/2\|_2 > a - \varepsilon.$$

We get that (g_n) is fundamental due to

$$\|(g_k - g_m)\|_2^2 = 2\|g_k\|_2^2 + 2\|g_m\|_2^2 - \|(g_k + g_m)\|_2^2 \leq 4(a + \frac{1}{n})^2 - 4(a - \varepsilon)^2,$$

and therefore this sequence converges.
Proof of lemma 10.18: We truncate f_n and denote truncated functions by $f_n^{(i)} = f_n \cdot I_{\{|f_n| \leq i\}}$, $n = 1, 2, ...$. Then for every i we get a bounded sequence $(f_n^{(i)})_{n \geq 1}$ for which the Komlos lemma can be applied in the following manner.

Problem 10.8 For each $n = 1, 2, ...$ there exist convex weights $(\lambda_n^n, ..., \lambda_{N_n}^n)$ such that $\sum_{j=n}^{N_n} \lambda_j^n f_j^{(i)}$ converge on space $L^2(\Omega)$ for each $i = 1, 2, ...$.
Hint: Use the Komlos lemma to find convex weights $\lambda_n^n, ..., \lambda_{N_n}^n$ with the convergent sum $\sum_{j=n}^{N_n} \lambda_j^n f_j^{(i)}$ as $n \to \infty$ for $i = 1, 2, ..., m$ and apply a diagonalization procedure to get a desirable result.

Now using uniform integrability of (f_n) we obtain that $f_n^{(i)}$ converges uniformly in n to f_n in $L^1(\Omega)$ as $i \to \infty$, and therefore uniformly in n:

$$\sum_{j=n}^{N_n} \lambda_j^n f_j^{(i)} \to \sum_{j=n}^{N_n} \lambda_j^n f_j \quad \text{in} \quad L^1(\Omega) \quad \text{as} \quad i \to \infty.$$

Lemma 10.19 *The sequence $(M_1^n)_{n \geq 1}$ is uniformly integrable.*

Proof Without loss of generality we can put $X_1 = 0$ and $X_t \leq 0$ for $t \leq 1$, because there is an obvious transition from X_t to $X_t - \mathbf{E}(X_1|F_t)$. In such a case $M_1^n = -A_1^n$ and
$$X_\tau^n = -\mathbf{E}(A_1^n|F_\tau) + A_\tau^n \tag{10.48}$$

for every s.t. τ w.r.t. $(F_t)_{t \in D_n}$.

Let us derive from assumption $X \in (D)$ that $(A_1^n)_{n \geq 1}$ is uniformly integrable. Define the following s.t.
$$\tau_n(c) = \inf\{(j-1)/2^n : A^n_{\frac{j}{2^n}} > c\} \wedge 1, \quad c > 0,$$

and note
$$\{\tau_n(c) < 1\} \subseteq \{\tau_n(\tfrac{c}{2}) < 1\}, \quad A^n_{\tau_n(c)} \leq c.$$

It follows from (10.48) that
$$X_{\tau_n(c)} \leq -\mathbf{E}(A_1^n|F_{\tau_n(c)}) + c.$$

Thus,
$$\begin{aligned}\mathbf{E}A_1^n \cdot I_{\{A_1^n > c\}} &= \mathbf{E}\mathbf{E}(A_1^n|F_{\tau_n(c)}) \cdot I_{\{\tau_n(c)<1\}} \\ &\leq c \cdot P(\tau_n(c) < 1) - \mathbf{E}X_{\tau_n(c)} \cdot I_{\{\tau_n(c)<1\}}\end{aligned}$$

and by (10.48)
$$\begin{aligned}\mathbf{E}(-X_{\tau_n(\frac{c}{2})}) \cdot I_{\{\tau_n(\frac{c}{2})<1\}} &= \mathbf{E}(A_1^n - A^n_{\tau_n(\frac{c}{2})}) \cdot I_{\{\tau_n(\frac{c}{2})<1\}} \\ &\geq \mathbf{E}(A_1^n - A^n_{\tau_n(\frac{c}{2})}) \cdot I_{\{\tau_n(c)<1\}} \\ &\geq \frac{c}{2} P\{\tau_n(c) < 1\}.\end{aligned}$$

These inequalities imply that
$$\mathbf{E}A_1^n \cdot I_{\{A_1^n > c\}} \leq -2\mathbf{E}X^n_{\tau_n(\frac{c}{2})} \cdot I_{\{\tau_n(c)<1\}} - \mathbf{E}X^n_{\tau_n(c)} \cdot I_{\{\tau_n(c)<1\}}. \tag{10.49}$$

We also note
$$P\{\tau_n(c) < 1\} = P\{A_1^n > c\} \leq \frac{1}{c}\mathbf{E}A_1^n = -\frac{1}{c}\mathbf{E}(M_1^n) = -\frac{1}{c}\mathbf{E}X_0,$$

and conclude that
$$\lim_{c \to \infty} P(\tau_n(c) < 1) = 0 \quad \text{uniformly in } n.$$

Assumption $X \in (D)$ together with (10.49) imply that $(A_1^n)_{n \geq 1}$ is uniformly integrable as well as $(M_1^n)_{n \geq 1} = (X_1 - A_1^n)_{n \geq 1}$.

Now we are ready for a limiting transition as $n \to \infty$ in (10.47).

First we extend M^n to the whole interval $[0, 1]$ by setting $M_t^n = \mathbf{E}(M_1^n|F_t)$. According to Lemma 10.18 and Lemma 10.19 there exist $M \in L_1(\Omega)$ and convex weights

10.4 The Doob-Meyer decomposition: proof and related remarks

$(\lambda_j^n)_{j=n,\ldots,N_n}$ such that $M_1^n(\lambda) = \sum_{j=n}^{N_n} \lambda_j^n M_1^j$ converges to M on space $L^1(\Omega)$. Moreover, by Jensen's inequality for each $t \in [0, 1]$:

$$M_t^n(\lambda) \to M_t = \mathbf{E}(M_1|F_t) \quad \text{as} \quad n \to \infty.$$

Secondly, we extend A^n to $[0, 1]$ by setting $A^n = \sum_{t \in D_n} A_t^n \cdot I_{(t-\frac{1}{2^n}, t]}$, $t \in [0, 1]$, $n = 1, 2, \ldots$, and denote $A^n(\lambda) = \sum_{j=n}^{N_n} \lambda_j^n A^j$.
Then the cadlag process

$$A_t = X_t - M_t, \quad \text{where} \quad t \in [0, 1],$$

satisfies the following relations

$$A_t^n(\lambda) = (X_t - M_t^n(\lambda)) \to (X_t - M_t) = A_t \quad \text{in} \quad L^1(\Omega) \quad \text{for} \quad t \in D, n \to \infty.$$

So, we find a subsequence $A_t^{n_k}(\lambda) \to A_t$ (a.s.) which we identify with $A_t^n(\lambda) \to A_t$ (a.s.). These standard considerations lead us to conclude that A is (a.s.) increasing on D, and therefore on the whole interval $[0, 1]$ due to its right-continuity. □

Lemma 10.20 *Process A is predictable.*

Proof: In fact we can restrict ourself by the proof that for all $t \in [0, 1]$ (a.s.)

$$\limsup_n A_t^n(\lambda) = A_t, \tag{10.50}$$

because the process A^n and $A^n(\lambda)$ are left-continuous and adapted, and hence they are predictable.

Problem 10.9 Let f_n and f be increasing functions form $[0, 1]$ to R^1 such that f is right-continuous and $f(t) = \lim_{n \to \infty} f_n(t)$, $t \in D$. Prove that

$$\limsup_n f_n(t) \le f(t) \quad \text{for all} \quad t \in [0, 1], \tag{10.51}$$

$$\lim_n f_n(t) = f(t), \quad \text{if } f \text{ is continuous at} \quad t \in [0, 1]. \tag{10.52}$$

Hint: By the right-continuity ($t_k \in D, t_k > t$):

$$f(t) = \lim_{k \to \infty} f(t_k) = \lim_{k \to \infty} (\lim_{n \to \infty} f_n(t_k)) \ge \limsup_n f_n(t),$$

since $\lim_n \sup f_n(t_k) \ge \lim_n \sup f_n(t)$.
Due to (10.51) and (10.52) the relation (10.50) can be broken at discontinuity points of A only. The process A is cadlag, and as was shown in Section 10.1 the points of jumps of A can be exhausted by a countable sequence of stopping times (see (10.2) and related remarks). So, it is sufficient for (10.50) that $\lim_n \sup A_\tau^n(\lambda) = A_\tau$ for all s.t. τ.
By (10.51) $\lim_\tau^n(\lambda) \le A_\tau$ and $A_\tau^n(\lambda) \le A_1^n(\lambda)$ which converges to A, on space $L^1(\Omega)$. Hence, by Fatou's lemma

$$\liminf_n \mathbf{E}A_\tau^n \leq \limsup_n \mathbf{E}A_\tau^n(\lambda) \leq \mathbf{E}\limsup_n A_\tau^n(\lambda) \leq \mathbf{E}A_\tau.$$

To prove that $\lim_n \mathbf{E}A_\tau^n = \mathbf{E}A_\tau$ we set $\sigma_n = \inf\{t \in D_n : t \geq \tau\}$. Then $\sigma_n \downarrow \tau$ as $n \to \infty$ and $A_\tau^n = A_{\sigma_n}^n$.
The process X belongs to class (D) and therefore

$$\mathbf{E}A_\tau^n = \mathbf{A}_{\sigma_n}^n = \mathbf{E}X_{\sigma_n} - \mathbf{E}M_0 \to \mathbf{E}X_\tau - \mathbf{E}M_0 = \mathbf{E}A_\tau.$$

Lemma 10.20 is proved and hence the decomposition (10.44) is true with a martingale M and an increasing predictable process A.

Chapter 11
General theory of stochastic processes in applications

Abstract The main goal of this chapter is to show how the general theory developed before can be applied to mathematical finance and statistics of random processes. In the area of mathematical finance a semimartingale financial market model is introduced. Applying to this general model the technique of stochastic exponents the fundamental questions of arbitrage and completeness of such a market are studied. These results have a number of corollaries for modeling and option pricing (Black-Scholes model and formula, Cox-Ross-Rubinstein model and formula etc). In the area of statistics of random processes the technique developed above gives a possibility to introduce semimartingale models. It is shown that classical discrete time and continuous time models of stochastic approximation are embedded in a semimartingale scheme. Moreover, it is proved that semimartingale stochastic approximation procedures are strong consistent and asymptotically normal under very wide conditions. In case of semimartingale regression the structural least-squared estimates are strong consistent and their sequential versions satisfy the important fixed accuracy property (see [3], [4], [11], [13], [18], [23], [30], [31], [32], [34], and [43]).

11.1 Stochastic mathematical finance

Suppose that besides the original measure on a stochastic basis $(\Omega, \mathcal{F}, (\mathcal{F}_t)_{t \geq 0}, P)$, we are given a measure $\tilde{P}$ locally equivalent to P and with local density $(Z_t)_{t \geq 0}$. The equivalence implies strict positivity of Z_t (P-a.s.), $t \geq 0$. Therefore, one can define a local martingale $N = (N_t)_{t \geq 0}$ with respect to P as a stochastic integral to P as a stochastic integral $N_t = \int_0^t Z_{s-}^{-1} dZ_s$. It leads to a stochastic exponent form of $Z_t = \mathcal{E}_t(N)$. We already know that a local martingale is transformed to a semimartingale (not a local martingale) under an equivalent change of measure. The next lemma contains a result of this type. To formulate it we denote $\mathcal{M}_{loc}(P)$ and $\mathcal{M}_{loc}(\tilde{P})$ classes of local martingales w.r. to P and $\tilde{P}$ correspondingly.

Lemma 11.1 Let $X = (X_t)_{t \geq 0}$ be a semimartingale on $(\Omega, \mathcal{F}, (\mathcal{F}_t)_{t \geq 0}, P)$, then

$$XZ \in \mathcal{M}_{loc}(P) \Rightarrow X \in \mathcal{M}_{loc}(\tilde{P}). \tag{11.1}$$

The proof of Lemma 11.1 can be given using a standard scheme. First, it is proved for martingales. Second, using localization one can extend it to local martingales. The first step is given with the help of change of probability in conditional expectations, which was stated before for a discrete time. A continuous time version of such change of measure is proved in the same way.

In fact, we prove (11.1) below in the framework of a semimartingale model of financial markets. We define a (B, S)-market as a collection of two positive semimartingales B and S on given stochastic basis $(\Omega, \mathcal{F}, (\mathcal{F}_t)_{t \geq 0}, P)$. The values of B_t and S_t are interpreted as the prices of a *non-risky asset* B and a *risky asset* S. A pair $\pi_t = (\beta_t, \gamma_t)_{t \geq 0}$ of stochastic processes with a predictable second component is defined as a *portfolio* of investment *strategy*. The portfolio has a *capital* $X_t^\pi = X_t^\pi(x) = \beta_t B_t + \gamma_t S_t$, where $X_0^\pi = \beta_0 B_0 + \gamma_0 S_0 = x \in \mathbb{R}$, $t \geq 0$.

The ratio $X_t = \frac{S_t}{B_t}$ is called the discounted price of S, and the ratio $\frac{X_t^\pi}{B_t}$ is called the discounted capital of the strategy π at the time $t \geq 0$.

In the set of all strategies we distinguish those portfolios π such that

$$\frac{X_t^\pi}{B_t} = \frac{X_0^\pi}{B_0} + \int_0^t \gamma_u d\left(\frac{S}{B}\right)_u. \tag{11.2}$$

We call them *self-financing* and denote their collection SF. We say that (B, S)-market admits an *arbitrage* at time $T > 0$ if there exists $\tilde{\pi} \in SF$ such that $X_0^{\tilde{\pi}} = 0$, $X_t^{\tilde{\pi}} \geq 0$, $t \leq T$, P-a.s., and $P(\omega : X_T^{\tilde{\pi}}(\omega) > 0) > 0$. Such a strategy $\tilde{\pi}$ is called the *arbitrage strategy*. Any probability measure $\tilde{P}$ locally equivalent to P is called a *local martingale measure*, if the discounted price $\left(\frac{S_t}{B_t}\right) \in \mathcal{M}_{loc}(\tilde{P})$.

Denote the set of such measures $\mathcal{M}(X, P)$. It is well-known that for a semimartingale (B, S)-market the *absence of arbitrage* is characterized as $\mathcal{M}(X, P) \neq \emptyset$.

By a *contingent claim* with exercise time T, we understand any non-negtaive $\mathcal{F}_T$-measurable random variable f. We remark that if such a claim f represents a pay-off of an option, then the option is called a *European option*. A strategy $\pi \in SF$ is said to be a *hedge* (hedging strategy) for f (for the option with the pay-off f) if $X_T^\pi(x) \geq f$ (a.s.) for some initial capital x. If in some class of hedging strategies π, there is a strategy π^* such that $X_t^{\pi^*} \leq X_t^\pi$ (a.s.) for all $t \in [0, T]$, then π^* is called a *minimal hedge* (in this class). Usually, the minimal hedge coincides with a *replicating strategy* π^* for which $X_T^{\pi^*} = f$ (a.s.). In this case, f is called *attainable*. The (B, S)-market is said to be *complete* if any contingent claim is attainable. For many financial markets, this notion is equivalent to the *uniqueness* of a local martingale measure P^*, i.e. $\mathcal{M}(X, P) = \{P^*\}$. Otherwise, the market is called *incomplete*. Correspondingly, pricing of an option with pay-off f in complete and incomplete markets is achieved with the help of martingale measures as follows:

11.1 Stochastic mathematical finance

$$\mathbb{C}(f) = \mathbf{E}^* \frac{f}{B_T} \text{ and } \mathbb{C}^*(f) = \sup_{\tilde{P} \in \mathcal{M}(X,P)} \tilde{\mathbf{E}} \frac{f}{B}, \qquad (11.3)$$

where $\mathbb{C}(f)$ is the *fair price* and $\mathbb{C}^*(f)$ is the so-called *upper price* of such an option.

So, we can see that an investigation of conditions under which $\tilde{P} \in \mathcal{M}(X,P)$ is extremely important for Mathematical Finance. Let us show how it can be done for a semimartingale (B,S)-market.

Suppose that a (B,S)-market is determined by the two equations

$$\begin{aligned} B_t &= B_0 + \int_0^t B_{u-} dh_u, \quad \Delta h_u > -1, \\ S_t &= S_0 + \int_0^t S_{u-} dH_u, \quad \Delta H_u > -1, \end{aligned} \qquad (11.4)$$

where h and H are given semimartingales.

Using stochastic exponents we can rewrite (11.4) as follows

$$B_t = B_0 \mathcal{E}_t(h) \text{ and } S_t = S_0 \mathcal{E}_t(H).$$

The problem in the model (11.4) is to determine conditions under which a measure P^* equivalent to P takes the process $X = \frac{S}{B}$ into a local martingale, i.e.

$$P^* \in \mathcal{M}(X,P) \Leftrightarrow X = \frac{S}{B} \in \mathcal{M}_{loc}(P^*). \qquad (11.5)$$

Let us check first when measure P is a martingale one for the market (11.4). Using the properties of stochastic exponents we have

$$\begin{aligned} X_t &= X_0 \mathcal{E}_t(H) \mathcal{E}_t^{-1}(h) = X_0 \mathcal{E}_t(H) \mathcal{E}_t(-h^*) \\ &= X_0 \mathcal{E}\left(H - h + \langle h^c, h^c \rangle + \sum \frac{(\Delta h)^2}{1+\Delta h} - \langle H^c, h^c \rangle - \sum \frac{\Delta H \Delta h}{1+\Delta h} \right) \\ &= X_0 \mathcal{E}_t \left(H - h + \langle h^c, h^c - H^c \rangle + \sum \frac{\Delta h(\Delta h - \Delta H)}{1+\Delta h} \right). \end{aligned} \qquad (11.6)$$

Denote

$$\Psi_t(h,H) = H_T - h_t + \langle h^c, h^c - H^c \rangle_t + \sum_{s \leq t} \frac{\Delta h(\Delta h - \Delta H)}{1+\Delta h}$$

and rewrite (11.6) in the form of the following stochastic differential equation

$$X_t = X_0 + \int_0^t X_{s-} d\Psi_s(h,H). \qquad (11.7)$$

Hence, $X \in \mathcal{M}_{loc}(P)$ if $\Psi(h, H) \in \mathcal{M}_{loc}(P)$. These considerations generate the idea how to find conditions to provide that $X \in \mathcal{M}_{loc}(P^*)$ for a local martingale measure P^* with the local density

$$Z_t = \frac{dP_t^*}{dP_t} = \mathcal{E}_t(N), \quad N_t = \int_0^t Z_{s-}^{-1} dZ_s \in \mathcal{M}_{loc}(P).$$

We just need to recognize when $XZ = X\mathcal{E}(N) \in \mathcal{M}_{loc}(P)$, and apply (11.1) of Lemma 11.1.

In view (11.6)-(11.7) we have

$$X_t \mathcal{E}_t(N) = X_0 \mathcal{E}_t(\Psi(h, H)) \mathcal{E}_t(N),$$

and by the properties of stochastic exponents obtain that

$$X_t \mathcal{E}_t(N) = X_0 \mathcal{E}_t(\Psi(h, H, N)), \qquad (11.8)$$

where

$$\Psi_t(h, H, N) = H_t - h_t + N_t + \langle (h - N)^c, (h - H)^c \rangle_t + \sum_{s \leq t} \frac{(\Delta h_s - \Delta N_s)(\Delta h_s - \Delta H_s)}{1 + \Delta h_s}.$$

Using relation (11.8), we arrive to the following theorem.

Theorem 11.1 *Let $X = \frac{S}{B}$ in the model (11.4). Then the following claims are true*
(1) If $\Psi(h, H) \in \mathcal{M}_{loc}(P)$, then $X \in \mathcal{M}_{loc}(P)$,
(2) If $\Psi(h, H, N) \in \mathcal{M}_{loc}(P)$, then $X \in \mathcal{M}_{loc}(P^*)$.

The above theorem presents a convenient methodology of finding of martingale measures for the model (11.4). Let us demonstrate this methodology for several partial cases of (11.4).

Example 11.1 The Black-Scholes model:

$$\begin{aligned} dB_t &= rB_t dt, \ B_0 = 1, \\ dS_t &= S_t(\mu dt + \sigma dW_t), \ S_0 > 0, \end{aligned} \qquad (11.9)$$

where $r, \mu \in \mathbb{R}_+, \sigma > 0, W = (W_t)_{t \geq 0}$ is a Wiener process.

In the model (11.9), $h_t = rt$, $H_t = \mu t + \sigma W_t$. Due to W is the sole source of randomness, we take N in the form $N_t = \phi W_t$. To use Theorem 11.1 we find that

$$\begin{aligned} \Psi(h, H, N) &= \mu t + \sigma W_t - rt + \phi W_t + \phi \sigma t \\ &= (\mu - r + \phi \sigma)t + (\sigma + \phi)W_t. \end{aligned}$$

Therefore, the condition (2) of Theorem 11.1 is satisfied if $\mu - r + \phi\sigma = 0$ or $\phi = -\frac{\mu-r}{\sigma}$, and we can construct the local density

$$Z_t = \frac{dP_t^*}{dP_t} = \mathcal{E}_t(N) = \exp\left\{-\frac{\mu - r}{\sigma} W_t - \frac{1}{2}\left(\frac{\mu - r}{\sigma}\right)^2 t\right\}. \qquad (11.10)$$

11.1 Stochastic mathematical finance

Using the Girsanov exponent (11.10) we find the martingale measure P^* under which $W_t^* = W_t + \frac{\mu-r}{\sigma}t$ will be a Wiener process by the Girsanov theorem. So, if $f = (S_T - K)^+$ is a contingent claim, then its initial price $\mathbb{C}(f)$ can be calculated as follows

$$\mathbb{C}(f) = \mathbf{E}^* \frac{(S_T - K)^+}{e^{rT}} = S_0 \Phi(d_+) - Ke^{-rT}\Phi(d_-), \tag{11.11}$$

$d_\pm = \frac{\ln(S_0/K)+(r\pm\sigma^2/2)T}{\sigma\sqrt{T}}$, and we again arrive to the Black-Scholes formula.

Example 11.2 The Merton model:

$$\begin{aligned} dB_t &= rB_t dt, \; B_0 = 1, \\ dS_t &= S_{t-}(\mu dt - \nu d\Pi_t), \end{aligned} \tag{11.12}$$

where $r, \mu \in \mathbb{R}_+$, $\nu < 1$, $\Pi = (\Pi_t)_{t \geq 0}$ is a Poisson process with parameter $\lambda > 0$.

In the model (11.12) $h_t = rt$, $H_t = \mu t - \nu\Pi_t$, and the martingale N_t will be chosen as $N_t = \psi(\Pi_t - \lambda t)$. Further,

$$\begin{aligned} \Psi(h, H, N) &= \mu t - \nu\Pi_t - rt + \psi(\Pi_t - \lambda t) - \psi \nu \Pi_t \\ &= (\mu - r - \nu\lambda - \psi\nu\lambda)t - \nu(\Pi_t - \lambda t) + \psi(\Pi_t - \lambda t) - \psi\nu(\Pi_t - \lambda t). \end{aligned}$$

Hence, the condition (2) of Theorem 11.1 is fulfilled if $\mu - r - \nu\lambda - \psi\nu\lambda = 0$ or $\psi = \frac{\mu-r}{\nu\lambda} - 1$ is the unique solution. The uniqueness means that a martingale measure P^* is also unique, and its local density has the following exponential form

$$Z_t = \frac{dP_t^*}{dP_t} = \mathcal{E}_t(N) = \exp[(\lambda - \lambda^*)t + (\ln\lambda^* - \ln\lambda)\Pi_t], \tag{11.13}$$

where $\lambda^* = \frac{\mu-r}{\nu}$ is a parameter of Π_t under measure P^*.

Using (11.13) we calculate a call option price $\mathbb{C}$ in the model (11.12) as follows.

$$\mathbb{C} = \mathbf{E}^* \frac{(S_T - K)^+}{e^{rT}}.$$

It is clear that

$$\begin{aligned} \frac{S_T}{B_T} &= \frac{S_0}{B_0} \exp[-\nu\Pi_T + \nu\lambda^*T] \prod_{t \leq T}(1 - \nu\Delta\Pi_t)e^{\nu\Delta\Pi_t} \\ &= \frac{S_0}{B_0} \exp[\Pi_T \ln(1-\nu) + \nu\lambda^*T]. \end{aligned} \tag{11.14}$$

Using (11.14) we find that

$$\mathbb{C} = \mathbf{E}^* \left(\frac{S_T}{B_T} - \frac{K}{B_T} \right)^+$$
$$= \sum_{n=0}^{\infty} e^{-\lambda^* T} \frac{(\lambda^* T)^n}{n!} \left(S_0 e^{n \ln(1-\nu) + \nu \lambda^* T} - B_T^{-1} K \right)^+. \quad (11.15)$$

Denote

$$n_0 = \inf \left[n : S_0 e^{n \ln(1-\nu) + \nu \lambda^* T} \geq \frac{K}{B_T} \right]$$
$$= \left[\frac{\ln(K/S_0) - \mu T}{\ln(1-\nu)} \right],$$

$$\Psi_p(x, y) = \sum_{n=x}^{\infty} e^{-y} \frac{y^n}{n!},$$

and find from (11.15) that

$$\mathbb{C} = S_0 \sum_{n=n_0}^{\infty} e^{-\lambda^* T + \lambda^* \nu T} e^{n \ln(1-\nu)} \frac{(\lambda^* T)^n}{n!} - K e^{-rT} \Psi_p(n_0, \lambda^* T)$$
$$= S_0 \sum_{n=n_0}^{\infty} e^{-\lambda^*(\nu - 1)T} \frac{(\lambda^*(1-\nu)T)^n}{n!} - K e^{-rT} \Psi_p(n_0, \lambda^* T) \quad (11.16)$$
$$= S_0 \Psi_p(n_0, \lambda^*(1-\nu)T) - K e^{-rT} \Psi_p(n_0, \lambda^* T).$$

Formula (11.16) is called the *Merton formula* for the price of a call option in the model (11.12).

Putting together models (11.9) and (11.12) we get a *jump-diffusion market model*:

$$B_t = r B_t dt, \quad B_0 = 1,$$
$$dS_t = S_{t-}(\mu dt + \sigma dW_t - \nu d\Pi_t), \quad S_0 > 0. \quad (11.17)$$

To calculate process $\Psi(h, H, N)$ in this case we choose $N_t = \phi W_t + \psi \Pi_t$ and find that

$$\Psi_t(h, H, N) = (\mu - r - \nu \lambda + \phi \sigma - \psi \lambda)t + \text{martingale}.$$

Hence, $\mu - r + \phi \sigma - \lambda \nu(1 + \psi) = 0$ to make $\Psi(h, H, N)$ a martingale. But this equation has infinitely many solutions (ϕ, ψ), and therefore, the model (11.15) admits infinitely many martingale measures, i.e. the market (11.17) is *incomplete*.

Let us consider a discrete-time model, called a *Binomial market* or the *Cox-Ross-Rubinstein model*. We show that such a model is embedded in the model (11.4).

Example 11.3 The model we are talking represents a kind of Binomial random walk:

11.1 Stochastic mathematical finance

$$\Delta B_n = B_n - B_{n-1} = rB_{n-1}, \ B_0 > 0,$$
$$\Delta S_n = S_n - S_{n-1} = \rho_n S_{n-1}, \ S_0 > 0, \quad (11.18)$$

where $(\rho_n)_{n=1,2,\ldots}$ is a sequence of independent random variables taking two values $b > a$ with probabilities p and $q = 1 - p$, $p \in (0, 1)$.

Assume also that $-1 < a < r < b$. Putting $B_t = B_n$ on $[n, n+1)$ (we do the same with $S_t, \mathcal{F}_t, \ldots$), we transform (11.18) to the model (11.4). This standard procedure allows to apply the theory developed for the semimartingale model.

In the case under consideration

$$\Delta h_n = r, \ \Delta H_n = \rho_n, \ \Delta N_n = \psi_n(\rho_n - \mu), \ \mu = \mathbf{E}\rho_n,$$

and

$$(1+r)\Delta \Psi_n(h, H, N) = \Delta H_n - \Delta h_n + \Delta N_n + \Delta N_n \Delta H_n.$$

Hence, the martingality property Ψ_n means that

$$\mathbf{E}(\rho_n - r + \psi_n(\rho_n - \mu) + \psi_n(\rho_n - \mu)\rho_n | \mathcal{F}_{n-1}) = 0,$$

which leads to

$$\psi_n = -\frac{\mu - r}{\sigma^2},$$

where $\sigma^2 = \mathbf{Var}(\rho_n)$.

As a result, we can construct a local density of a martingale measure P^* here in the form of stochastic exponent

$$Z_n = \mathcal{E}_n\left(-\frac{\mu - r}{\sigma^2}\sum(\rho_n - \mu)\right). \quad (11.19)$$

As in previous examples we want to derive the price of a call option $(S_N - K)^+$, where $N \geq 1$. According to the general theory the price $\mathbb{C}$ can be calculated as

$$\mathbb{C} = \mathbf{E}^*(1+r)^{-N}(S_N - K)^+$$
$$= (1+r)^{-N}\mathbf{E}^*(S_N - K)^+ \quad (11.20)$$
$$= (1+r)^{-N}\mathbf{E}\mathcal{E}_N\left(-\frac{\mu - r}{\sigma^2}\sum(\rho_k - \mu)\right)(S_N - K)I_{\{S_N > K\}},$$

where we used (11.19).

Define $k_0 = \min\{k \leq N : S_0(1+b)^k(1+a)^{N-k} > K\}$ and find that $k_0 \ln(1+b) + (N-k_0)\ln(1+a) > \ln\frac{K}{S_0}$ or $k_0 = \left[\ln\frac{K}{S_0(1+a)^N} \bigg/ \ln\frac{1+b}{1+a}\right] + 1$, where $[x]$ is an integer part of $x \in \mathbb{R}$.

Denote $p^* = \frac{r-a}{b-a}$ and derive for the term with K in (11.20) using elementary equalities $\mu = p(b-a) + a$, $\sigma^2 = (b-a)^2 p(1-p)$ that

$$
\begin{aligned}
(1+r)^{-N} &\mathbf{E}\mathcal{E}_N\left(-\frac{\mu-r}{\sigma^2}\sum(\rho_k-\mu)\right)KI_{\{S_N>K\}}\\
=&K(1+r)^{-N}\sum_{k=k_0}^{N}\binom{N}{k}\left(1-\frac{\mu-r}{\sigma^2}(b-\mu)\right)^k p^k\left(1-\frac{\mu-r}{\sigma^2}(a-\mu)\right)^{N-k}(1-p)^{N-k}\\
=&K(1+r)^{-N}\sum_{k=k_0}^{N}\binom{N}{k}\left(\frac{p^*}{p}\right)^k p^k\left(\frac{1-p^*}{1-p}\right)^{N-k}(1-p)^{N-k}\\
=&K(1+r)^{-N}\sum_{k=k_0}^{N}\binom{N}{k}(p^*)^k(1-p^*)^{N-k}.
\end{aligned}
$$
(11.21)

To calculate the term with $S_N = S_0\mathcal{E}_N(\sum \rho_k)$ in (11.20) we use a multiplication rule of stochastic exponents:

$$
\begin{aligned}
(1+r)^{-N} &\mathbf{E}\mathcal{E}_N\left(-\frac{\mu-r}{\sigma^2}\sum(\rho_k-\mu)\right)S_N I_{\{S_N>K\}}\\
=&S_0(1+r)^{-N}\mathbf{E}\mathcal{E}_N\left(-\frac{\mu-r}{\sigma^2}\sum(\rho_k-\mu)+\sum\rho_k-\frac{\mu-r}{\sigma^2}\sum(\rho_k-\mu)\rho_k\right)I_{\{S_N>K\}}\\
=&S_0(1+r)^{-N}\sum_{k=k_0}^{N}\binom{N}{k}\left(1-\frac{\mu-r}{\sigma^2}(b-\mu)+b-\frac{\mu-r}{\sigma^2}(b-\mu)b\right)^k p^k\\
&\times\left(1-\frac{\mu-r}{\sigma^2}(a-\mu)+a-\frac{\mu-r}{\sigma^2}(a-\mu)a\right)^{N-k}(1-p)^{N-k}\\
=&S_0(1+r)^{-N}\sum_{k=k_0}^{N}\binom{N}{k}\left(\frac{p^*}{p}(1+b)\right)^k p^k\left(\frac{1-p^*}{1-p}(1+a)\right)^{N-k}(1-p)^{N-k}\\
=&S_0(1+r)^{-N}\sum_{k=k_0}^{N}\binom{N}{k}\left(p^*\frac{1+b}{1+r}\right)^k\left((1-p^*)\frac{1+a}{1+r}\right)^{N-k}.
\end{aligned}
$$
(11.22)

Introducing the notations $\tilde{p} = \frac{1+b}{1+r}p^*$ and $\mathbb{B}(j,N,p) = \sum_{k=j}^{N}\binom{N}{k}p^k(1-p)^{N-k}$ and putting together (11.21)-(11.22) we get from (11.20) the *Cos-Ross-Rubinstein formula* for the initial price of a call option

$$\mathbb{C} = S_0\mathbb{B}(k_0, N, \tilde{p}) - K(1+r)^{-N}\mathbb{B}(k_0, N, p^*).$$

11.2 Stochastic Regression Analysis

Proposed and developed below an extension of classical regression models and techniques is based on those theoretical findings that was delivered in the previous chapter, and can be called a *Stochastic Regression Analysis*.

11.2 Stochastic Regression Analysis

We start with the classical problem of *stochastic approximation*. It consists of a construction of a stochastic sequence θ_n or a stochastic process θ_t that converges in some probabilistic sense to unique root $\theta \in \mathbb{R}$ of the regression equation

$$R(\theta) = 0, \tag{11.23}$$

where R is a regression function.

In classical theory, a solution to (11.23) is given by the *Robbins-Monro procedure*

$$\theta_n = \theta_{n-1} - \gamma_n y_n, \quad n = 1, 2, \ldots, \tag{11.24}$$

where the sequence of observations y_n is such that

$$y_n = R(\theta_{n-1}) + \xi_n, \tag{11.25}$$

$(\xi_n)_{n=1,2,\ldots}$ is a sequence of independent random variables or martingale-differences, $(\gamma_n)_{n=1,2,\ldots}$ is a numerical positive sequence converging to zero.

The convergence (a.s.) of the procedure (11.24)-(11.25) for a continuous linearly bounded regression function $R(x)$ such that

$$R(x)(x - \theta) > 0 \text{ for all } x \neq \theta$$

is guaranteed by finiteness of the variance (conditional variance) of observation errors ξ_n and the following conditions:

$$\sum_{n=1}^{\infty} \gamma_n = \infty, \tag{11.26}$$

$$\sum_{n=1}^{\infty} \gamma_n^2 < \infty. \tag{11.27}$$

Diffusion analogues of (11.24)-(11.25) and (11.26)-(11.27) are

$$d\theta_t = -\gamma_t R(\theta_t) dt - \gamma_t dW_t,, \tag{11.28}$$

$$\int_0^{\infty} \gamma_s ds = \infty,, \tag{11.29}$$

$$\int_0^{\infty} \gamma_s^2 ds < \infty, \tag{11.30}$$

where W_t is a Wiener process and γ_t is a positive deterministic function tending to zero as $t \to \infty$.

Under conditions (11.29)-(11.30) the procedure θ_t converges to θ as $t \to \infty$.

The leading idea to generalize (11.24)-(11.25) and (11.28) consists in the possibility of describing such stochastic algorithms as strong solutions of some special classes of stochastic differential equations with respect to semimartingales. It leads to a generalized Robbins-Monro procedure as a process θ_t satisfying the stochastic differential equation

$$\theta_t = \theta_0 - \int_0^t \gamma_s R(\theta_{s-}) da_s - \int_0^t \gamma_s dm_s, \qquad (11.31)$$

where θ_0 is a finite $\mathcal{F}_0$-measurable random variable, a predictable process $a \in \mathcal{A}_{loc}^+$, $m \in \mathcal{M}_{loc}^2$, and γ is a predictable process decreasing to zero (a.s.) as $t \to \infty$.

To simplify the demonstration how martingale methods work here we consider a linear case only: $R(x) = \beta(x - \theta), \beta > 0$.

In this case (11.31) is reduced to

$$\theta_t - \theta = \theta_0 - \theta - \int_0^t \gamma_s \beta(\theta_{s-} - \theta) da_s - \int_0^t \gamma_s dm_s. \qquad (11.32)$$

Let us assume that (a.s.)

$$\int_0^\infty \gamma_s da_s = \infty, \qquad (11.33)$$

$$\int_0^\infty \gamma_s^2 d\langle m, m \rangle_s < \infty. \qquad (11.34)$$

Define the following stochastic exponent

$$\mathcal{E}_t(-\beta\gamma \cdot a) = \mathcal{E}_t\left(-\beta \int \gamma_s da_s\right),$$

and assume that $\beta\gamma_t \Delta a_t < 1$ to provide that $\mathcal{E}_t(-\beta\gamma \cdot a) > 0$ (a.s.).

Applying the Ito formula to $\mathcal{E}_t^{-1}(-\beta\gamma \cdot a)(\theta_t - \theta)$ we arrive from (11.32) to

$$\theta_t - \theta = \mathcal{E}_t(-\beta\gamma \cdot a)(\theta_0 - \theta) - \mathcal{E}_t(-\beta\gamma \cdot a) \int_0^t \gamma_s \mathcal{E}_s^{-1}(-\beta\gamma \cdot a) dm_s. \qquad (11.35)$$

The first term in the right-hand side of (11.35) converges to zero (a.s.) as $t \to \infty$ because of $\mathcal{E}_t(-\beta\gamma \cdot a) \to 0$ (a.s.), $t \to \infty$. Second term in (11.35) is treated by the arguments of the Large Numbers Law for square integrable martingales. Using (11.34) we have that (a.s.)

$$\int_0^\infty \mathcal{E}_s^2(-\beta\gamma \cdot a) \gamma_s^2 \mathcal{E}_s^2(-\beta\gamma \cdot a) d\langle m, m \rangle_s = \int_0^\infty \gamma_s^2 d\langle m, m \rangle_s < \infty.$$

Hence, the second term of (11.35) converges to zero (a.s.) as $t \to \infty$. As a result, we get the convergence (a.s.) $\theta_t \to \theta$ as $t \to \infty$.

To investigate an asymptotic normality of the procedure (11.32) we simplify the situation assuming that a_t is a deterministic function and m_t is a Gaussian martingale satisfying conditions:

11.2 Stochastic Regression Analysis

$$\langle m, m \rangle_t = \sigma^2 a_t, \ \sigma^2 > 0, \ \gamma_t = \frac{\alpha}{1 + a_t}, \ \alpha > 0, \ \beta\alpha < 1,$$

$$a_t \uparrow, \ t \to \infty, \ \sum_{0 < s < \infty} \left(\frac{\Delta a_s}{1 + a_s}\right)^2 < \infty.$$

Under these conditions we have

$$\mathcal{E}_t(-\beta\gamma \cdot a) = \mathcal{E}_t(-\beta\alpha) \sim (1 + a_t)^{-\beta\alpha} \text{ as } t \to \infty.$$

Multiplying (11.35) by $(1 + a_t)^{1/2}$ we obtain

$$(1 + a_t)^{1/2}(\theta_t - \theta) = (1 + a_t)^{1/2} \mathcal{E}_t(-\beta\alpha)(\theta_0 - \theta)$$
$$- (1 + a_t)^{1/2} \mathcal{E}_t(-\beta\alpha) \int_0^t \frac{\alpha}{1 + a_s} \mathcal{E}_s^{-1}(-\beta\alpha) dm_s. \quad (11.36)$$

Let us note that

$$(1 + a_t)^{1/2} \mathcal{E}_t(-\beta\alpha) \sim (1 + a_t)^{1/2 - \beta\alpha} \text{ as } t \to \infty,$$

and therefore, under $2\beta\alpha > 1$ we obtain that the first term of (11.36) converges to zero (a.s.) as $t \to \infty$.

The second term of (11.36) has a Gaussian distribution, and hence, we need to calculate the asymptotic value of the variance of this term:

$$(1 + a_t) \mathcal{E}_t^2(-\beta\alpha) \int_0^t \mathcal{E}_s^2(-\beta\alpha) \frac{\alpha^2 \sigma^2}{1 + a_s} da_s \to \frac{\alpha^2 \sigma^2}{2\beta\alpha - 1} \text{ as } t \to \infty. \quad (11.37)$$

Finally, we can arrive to conclusion that

$$(1 + a_t)^{1/2}(\theta_t - \theta) \xrightarrow[t \to \infty]{d} N\left(0, \frac{\alpha^2 \sigma^2}{2\beta\alpha - 1}\right). \quad (11.38)$$

As a consequence of (11.38) we get the well-known classical results from the theory of stochastic approximation for *discrete* and *continuous* time (11.24)-(11.25) and (11.28):

$$n^{1/2}(\theta_n - \theta) \xrightarrow[n \to \infty]{d} N\left(0, \frac{\alpha^2 \sigma^2}{2\beta\alpha - 1}\right),$$

$$t^{1/2}(\theta_t - \theta) \xrightarrow[t \to \infty]{d} N\left(0, \frac{\alpha^2 \sigma^2}{2\beta\alpha - 1}\right).$$

Another important problem of regression analysis is the problem of estimation of the unknown parameter of a linear regression function.

Suppose we observe the process X_t having the following structure

$$X_t = \int_0^t f_s \, da_s \theta + M_t, \qquad (11.39)$$

where $a \in \mathcal{A}_{loc}^+$ and predictable, $M \in \mathcal{M}_{loc}^2$, a predictable function f_t such that, $\int_0^t f_s^2 \, da_s = F_t < \infty$ (a.s.), $t \geq 0$, $\theta \in \mathbb{R}$ is an unknown parameter.

Define the following structural Least Squares estimate as follows

$$\theta_t = F_t^{-1} \int_0^t f_s \, dX_s, \qquad (11.40)$$

where we assume that $F_t > 0$ (a.s.) and $F_t \in \mathcal{A}_{loc}^+$ and predictable.

Assuming that $F_t \to \infty$ (a.s.) as $t \to \infty$, we rewrite (11.40):

$$\begin{aligned}\theta_t &= F_t^{-1} \int_0^t f_s^2 \, da_s \theta + F_t^{-1} \int_0^t f_s \, dM_s \\ &= \theta + F_t^{-1} \int_0^t f_s \, dM_s.\end{aligned} \qquad (11.41)$$

Obviously, we can study the asymptotic behaviour of θ_t with the help of the Large Numbers Law. To apply such LNL we find from (11.41) that (a.s.)

$$\int_0^\infty F_s^{-2} \, d\left\langle \int_0^t f_s \, dM_s, \int_0^t f_s \, dM_s \right\rangle = \int_0^\infty F_s^{-2} f_s^2 \, d\langle M, M, \rangle_s < \infty. \qquad (11.42)$$

So, if $F_t \to \infty$ (a.s.), $t \to \infty$, and (11.42) is fulfilled then $\theta_t \to \theta$ (a.s.) $t \to \infty$.

Let us continue our study of estimates (11.40), by considering their *sequential analog*. To do this in the model (11.39) we assume that

$$\frac{d\langle M, M \rangle_t}{da_t} \leq \xi \gamma_t, \qquad (11.43)$$

$$\int_0^t \gamma_s^{-1} f_s^2 \, da_s \in \mathcal{A}_{loc}^+ \text{ and predictable},$$

where ξ is a positive r.v. and γ_t is a positive predictable process.

Next, for $H > 0$ we put

$$\tau_H = \inf\left\{ t : \int_0^t \gamma_s^{-1} f_s^2 \, da_s \geq H \right\}$$

with $\tau_H = \infty$ if the set in $\{\cdot\}$ is empty.

On the set $\{\tau_H < \infty\}$ we define a random variable β_H by the relation

$$\int_{(0,\tau_H)} \gamma_s^{-1} f_s^2 \, da_s + \beta_H \gamma_{\tau_H}^{-1} f_{\tau_H}^2 \Delta a_{\tau_H} = H, \qquad (11.44)$$

11.2 Stochastic Regression Analysis

and we put $\beta_H = 0$ on the set $\{\tau_H = \infty\}$. Then $\beta_H \in [0, 1)$ and it is a $\mathcal{F}_{\tau_H-}$-measurable random variable.

Let us define the following *sequential* Least Squares estimate

$$\hat{\theta}_H = H^{-1}\left[\int_{(0,\tau_H)} \gamma_s^{-1} f_s dX_s + \beta_H \gamma_{\tau_H}^{-1} f_{\tau_H} \Delta X_{\tau_H}\right]. \tag{11.45}$$

The next theorem shows a nice property of $\hat{\theta}_H$ called a *fixed accuracy*.

Theorem 11.2 *Suppose* (11.43)-(11.45) *are fulfilled,* $\mathbf{E}\xi < \infty$ *and* $\int_0^\infty \gamma_s^{-1} f_s^2 da_s = \infty$ *(a.s.). Then* $P\{\omega : \tau_H < \infty\} = 1$, $\mathbf{E}\hat{\theta}_H = \theta$, *and*

$$\mathbf{Var}\hat{\theta}_H \leq H^{-1}\mathbf{E}\xi. \tag{11.46}$$

Proof The finiteness (a.s.) of τ_H follows from the relation $\{\tau_H \leq T\} = \left\{\omega : \int_0^T \gamma_s^{-1} f_s^2 da_s \geq H\right\}$. Using (11.45) we find that

$$\hat{\theta}_H = \theta + H^{-1} N_{\tau_H}, \tag{11.47}$$

where $N_t = \int_0^t I_{(s<\tau_H)} \gamma_s^{-1} f_s dM_s + I_{\{t=\tau_H\}} \beta_H \gamma_{\tau_H}^{-1} f_{\tau_H} \Delta M_{\tau_H}$.

Since the process N_t is a stochastic integral with respect to $M \in \mathcal{M}_{loc}^2$, by the properties of stochastic integrals we obtain that

$$\langle N, N\rangle_t = \int_0^t I_{(s<\tau_H)} \gamma_s^{-1} f_s^2 d\langle M, M,\rangle_s + I_{\{t=\tau_H\}} \beta_H^2 \gamma_{\tau_H}^{-2} f_{\tau_H}^2 \Delta\langle M, M\rangle_{\tau_H} \tag{11.48}$$

Hence, by (11.44) we get

$$\langle N, N\rangle_{\tau_H} = \xi\left[\int_{(0,\tau_H)} \gamma_s^{-1} f_s^2 da_s + \beta_H \gamma_{\tau_H}^{-1} f_{\tau_H}^2 \Delta a_{\tau_H}\right] = \xi H,$$

and consequently, $N_{t\wedge\tau_H} \in \mathcal{M} \cap \mathcal{M}^2$ and

$$\mathbf{E}N_{\tau_H} = 0, \quad \mathbf{E}N_{\tau_H}^2 = H\mathbf{E}\xi,$$

which leads to (11.46).

Example 11.4 Let us consider the first order autoregression model:

$$x_t = \theta x_{t-1} + e_t, \; x_0 = 0, t = 1, 2, \ldots,$$

where $(e_t)_{t=1,2,\ldots}$ is a martingale-difference w.r. to $(\mathcal{F}_t)$ and $\mathbf{E}(e_t^2|\mathcal{F}_{t-1}) \leq \xi$, $\mathbf{E}\xi < \infty$. Then (11.43) is fulfilled with $\gamma_t = 1$, and the estimate $\hat{\theta}_H$ takes the form

$$\hat{\theta}_H = H^{-1}\left[\sum_{k=1}^{\tau_H-1} x_k x_{k-1} + \beta_H x_{\tau_H} x_{\tau_H-1}\right],$$

$$\tau_H = \inf\left\{n : \sum_{k=1}^{n} x_{k-1}^2 \geq H\right\},$$

$$\beta_H = x_{\tau_H}^{-2}\left[H - \sum_{k=1}^{\tau_H-1} x_k^2\right].$$

Embedding the present model in the general model (11.39) in a standard way, we get

$$\mathbf{E}\hat{\theta}_H = \theta, \quad \mathbf{Var}\hat{\theta}_H \leq H^{-1}\mathbf{E}\xi.$$

Chapter 12
Supplementary problems

Abstract The list below contains problems which are related to all chapters of the book. Some of them are numerical and some others are pure theoretical, but in any case they are helped for both students and instructors. Students can improve their understanding and scope. Instructors can transform most of the problems for teaching and examination purposes. The following references might be useful to create detailed solutions (see [1], [5], [7], [10], [11], [13], [14], [15], [16], [17], [18], [21], [22], [23], [24], [27], [28], [29], [30], [31], [35], [37], [43], [44], and [45]).

Problem 12.1 Let X be a standard normal random variable and Y be a Bernoulli random variable such that $P(Y = 1) = P(Y = -1) = \frac{1}{2}$, and X, Y are independent. Prove that
(a) $Z = YX \sim N(0, 1)$,
(b) Z and X are uncorrelated, but they are not independent.

Solution: The claim (a) is obvious. To prove (b) we calculate $cov(X, Z) = \mathbb{E}(XZ) = \mathbb{E}X^2 \mathbb{E}Y = 1 \cdot 0 = 0$. Further, $P(X \geq 1, Z \geq 1) = P(X \geq 1, Y = 1) = P(X \geq 1) \cdot P(Y = 1) = \frac{1}{2}P(X \geq 1)$. But $P(X \geq 1) \cdot P(Z \geq 1) = (P(X \geq 1))^2$ and $P(X \geq 1) \simeq 0.1587$, and, hence, $\frac{1}{2}P(X \geq 1) \neq (P(X \geq 1))^2$.

∎

Problem 12.2 Assume that stochastic process $X_t = e^{W_t}$, where $W = (W_t)_{t \geq 0}$ is a standard Wiener process. Calculate the drift and diffusion coefficients of process $X = (X_t)_{t \geq 0}$:

$$b(x) = \lim_{h \to 0} h^{-1} [E(X_{t+h} - X_t | X_t = x)],$$

$$\sigma^2(x) = \lim_{h \to 0} h^{-1} \left[E\left((X_{t+h} - X_t)^2 | X_t = x\right) \right]$$

for $t, x \in R_+^1$.

Hint: Use the fact that
$$W_{t+h} - W_t \sim N(0, \sqrt{h}), \quad h > 0$$

Problem 12.3 Let $X_i \sim N(\mu_i, \sigma_i^2), i = 1, 2,$ and $X_1 + X_2 \sim N(\mu_1 + \mu_2, \sigma_1^2 + \sigma_2^2 + 2\rho\sigma_1\sigma_2)$. Then $cov(X_1, X_2) = \rho\sigma_1\sigma_2$ and $corr(X_1, X_2) = \rho$.

Hint: Use the formula for bivariate normal distribution.

Problem 12.4 (Theorem of Slutsky)

Let r.v's $X_n \xrightarrow[n\to\infty]{d} X$ and the sequence of real numbers $a_n \to a \in \mathbb{R}^1$. Then $X_n + a_n \xrightarrow[n\to\infty]{d} X + a$ and $a_n X_n \xrightarrow[n\to\infty]{d} aX$

Solution:

Let us note for $\forall \varepsilon > 0, x \in \mathbb{R}^1$, that

$$P(X_n + a_n \leq x) = P(X_n + a_n \leq x, |a_n - a| < \varepsilon) + P(X_n + a_n \leq x, |a_n - a| \geq \varepsilon) \leq$$
$$\leq P(X_n \leq x - a + \varepsilon) + P(|a_n - a| \geq \varepsilon).$$

It follows from here for all points $x - a + \varepsilon$ of continuity of distribution function F_x that

$$\limsup_n F_{X_n + a_n}(x) \leq F_x(x - a + \varepsilon).$$

So, due to arbitrary choice of $\varepsilon > 0$ we derive

$$\lim_{n\to\infty} F_{X_n + a_n}(x) = F_X(x - a) = F_{X+a}(x)$$

for all x at which F_{X+a} is continuous.

The second claim is proved in the same way.

Problem 12.5 Let $X_n \sim N(\mu_n, \sigma_n^2)$ and $\mu_n \to \mu, \sigma_n^2 \to \sigma^2, n \to \infty$. Then $X_n \xrightarrow[n\to\infty]{d} X \sim N(\mu, \sigma^2)$.

Solution:

Denote $Z_n = \frac{X_n - \mu_n}{\sigma_n} \sim N(0, 1)$ and, hence, $Z_n \xrightarrow[n\to\infty]{d} Z \sim N(0, 1)$. Using Problem 12.4, we obtain that

$$X_n = \sigma_n Z_n + \mu_n \xrightarrow[n\to\infty]{d} X = \sigma Z + \mu.$$

Problem 12.6 (Borel-Cantelli lemma) Let $(A_n)_{n=1,2,\ldots}$ be a sequence of events, and $C = \bigcap_{m=1}^{\infty} \bigcup_{n=m}^{\infty} A_n$. Then

12 Supplementary problems

1. $P(C) = 0$, if $\sum_{n=1}^{\infty} P(A_n) < \infty$;

2. $P(C) = 1$, if $(A_n)_{n=1,2,...}$ are independent and $\sum_{n=1}^{\infty} P(A_n) = \infty$.

Solution: 1. We have $P(C) \leq P\left(\bigcup_{n=m}^{\infty} A_n\right) \leq \sum_{n=m}^{\infty} P(A_n)$ for each $n = 1, 2, ...$ Take $\varepsilon > 0$ and find N_ε big enough that $\sum_{n=N_\varepsilon}^{\infty} P(A_n) < \varepsilon$. Hence, for all $N > N_\varepsilon$ we obtain $P(C) < \varepsilon$.

2. Let us note that $P((\bigcup_{n=m}^{\infty} A_n)^c) \leq P(\bigcup_{n=m}^{\infty} A_n^c) \leq P(\bigcup_{n=m}^{m+M} A_n^c)$ for any $M > 0$.

But (A_n) are independent, and, hence, $P((\bigcup_{n=m}^{\infty} A_n)^c) \leq \prod_{n=m}^{m+M} (1 - P(A_n)) \leq \exp\left\{-\sum_{n=m}^{m+M} P(A_n)\right\}$, where we used the inequality $1 - x \leq e^{-x}$ for $x \in [0, 1]$. The claim follows from the inequality above for probabilities.

∎

Problem 12.7 Let $\mu = \sum_{n=1}^{\infty} \alpha_n P_n$, where (P_n) and (α_n) are sequences of probability measures and positive numbers respectively. Define $\nu = \sum_{n=1}^{\infty} \beta_n Q_n$, where (Q_n) and (β_n) are sequences of probability measures $Q_n \ll P_n$ and non-negative numbers. Prove that $\nu \ll \mu$.

Solution:

For any A with $\mu(A) = 0$ we have $\sum_{1}^{\infty} \alpha_n P_n = \mu(A) = 0$ and, hence, $P_n(A) = 0$ for all n. So, $Q_n(A) = 0$ and therefore $\nu(A) = \sum_{n=1}^{\infty} \beta_n Q_n = 0$ and we get $\nu \ll \mu$.

∎

Problem 12.8 Let $([0, 1], \mathcal{B}(0, 1), l)$ be a Borel space with the Lebesgue probability measure l, and X and Y be random variables: $X(\omega) = 2\omega^2$ and

$$Y(\omega) = \begin{cases} 0, & \omega \in [0, \frac{1}{3}] \\ 2, & \omega \in [\frac{1}{3}, \frac{2}{3}] \\ 1, & \omega \in [\frac{2}{3}, 1] \end{cases}.$$

Find $\mathbb{E}(X|Y)$.

Solution: For $\omega \in [0, \frac{1}{3}]$ we have $\mathbb{E}(X|Y)(\omega) = \dfrac{\int_0^{\frac{1}{3}} x dP}{P([0, \frac{1}{3}])} = \dfrac{1}{\frac{1}{3}} \int_0^{\frac{1}{3}} 2\omega^2 d\omega = \frac{2}{27}$. Values of $\mathbb{E}(X|Y)$ on other sets $[\frac{1}{3}, \frac{2}{3}]$ and $[\frac{2}{3}, 1]$ can be determined in the same way: $\frac{14}{27}$ and $\frac{38}{27}$ correspondingly.

∎

Problem 12.9 Let $(\varepsilon_n)_{n=1,2,...,N}$ be a sequence of independent random variables with values $+1$ and -1 taking with probability $\frac{1}{2}$. Define $X_n = (-1)^n \cos\left(\pi \sum_{k=1}^{n} \varepsilon_k\right)$, $n = 1, 2, ..., N$ and prove that $(X_n)_{n=1...N}$ is a martingale with respect to a natural filtration $\mathcal{F}_n = \mathcal{F}_n^\varepsilon = \mathcal{F}_n^X$.

Solution: We represent the sequence (X_n) as follows: $X_n = (-1)^n \frac{1}{2} \left[e^{i\pi \sum_1^n \varepsilon_k} + e^{-i\pi \sum_1^n \varepsilon_k} \right]$, $i = \sqrt{-1}$.

Using independence of $(\varepsilon_n)_{1...N}$ we have $\mathbb{E}(X_n|\mathcal{F}_{n-1}) = \frac{1}{2}(-1)^n \left[\mathbb{E}e^{i\pi\varepsilon_n} \cdot e^{i\pi \sum_1^{n-1} \varepsilon_k} + \mathbb{E}e^{-i\pi\varepsilon_n} \cdot e^{-i\pi \sum_1^{n-1} \varepsilon_k} \right]$.

Further, applying an obvious relation $\mathbb{E}e^{i\pi\varepsilon_n} = \frac{1}{2}(e^{i\pi} + e^{-i\pi}) = -1$ we obtain

$$\mathbb{E}(X_n|\mathcal{F}_{n-1}) = \frac{(-1)^{n-1}}{2} \left[e^{i\pi \sum_1^{n-1} \varepsilon_k} + e^{-i\pi \sum_1^{n-1} \varepsilon_k} \right] = (-1)^{n-1} \cos\left(\pi \sum_1^{n-1} \varepsilon_k\right) = X_{n-1}.$$

∎

Problem 12.10 Let values and joint distribution of random variables X and Y are given in the table

X \ Y	-0.1	0	0.1
-0.2	0.1	0	0.4
0.1	0.3	0.1	0.1

Find marginal distributions of X and Y, average of Y and $\mathbb{E}(Y|X)$.

Solution: We have from the table above that

$$P(X = -0.2) = 0.1 + 0.4 = 0.5,$$
$$P(X = 0.1) = 0.3 + 0.1 + 0.1 = 0.5,$$
$$P(Y = -0.1) = 0.1 + 0.3 = 0.4,$$
$$P(Y = 0) = 0.1, \quad P(Y = 0.1) = 0.4 + 0.1 = 0.5$$

which give us marginal distributions. We also derive from the above equalities that $\mathbb{E}Y = -0.1 \cdot 0.4 + 0.1 \cdot 0.5 = 0.01$.

To calculate the conditional expectation $\mathbb{E}(Y|X)$ we write

$$\mathbb{E}(Y|X) = \mathbb{E}(Y|X = -0.2) \cdot I_{\{X=-0.2\}} + \mathbb{E}(Y|X = 0.1) \cdot I_{\{X=0.1\}}.$$

Calculating

$$\mathbb{E}(Y|X = -0.2) = -0.1 \cdot P(Y = -0.1|X = -0.2) + 0.1 \cdot P(Y = 0.1|X = -0.2) =$$
$$= \frac{-0.1 \cdot 0.1 + 0.1 \cdot 0.4}{0.5} = 0.06,$$

and similarly

$$\mathbb{E}(Y|X=0.1) = \frac{-0.1 \cdot 0.3 + 0.1 \cdot 0.1}{0.5} = -0.04,$$

we obtain

$$E(Y|X) = 0.06 \cdot I_{\{X=-0.2\}} - 0.04 \cdot I_{\{X=0.1\}}.$$

∎

Problem 12.11 Let $(X_n)_{n=1,2,\cdots}$ be a sequence of independent random variables such that $P(X_n = 1) = p$, $P(X_n = -1) = 1 - p$, $1 < p < \frac{1}{2}$. Show that the following stochastic sequences are martingales with respect to a natural filtration $(\mathcal{F}_n)_{n=1,2,\cdots}$ generated by (X_n).
(a) $M_n = \sum\limits_{k=1}^{n} X_k - n \cdot (2p - 1)$;
(b) $Y_n = M_n^2 - 4np(1 - p)$;
(c) $Z_n = (\frac{1-p}{p})^{\sum\limits_{1}^{n} X_k}$.

Hint: Check the martingale property.

∎

Problem 12.12 Let X_0 be a random variable such that $P(X_0 = 2) = P(X_0 = 0) = \frac{1}{2}$. Define $X_n = n \cdot X_{n-1}$, $n = 1, 2, \cdots$ and $M_n = X_n - \mathbb{E}X_n$. Prove that (M_n) is not a martingale with respect to the natural filtration $(\mathcal{F}_n)$.

Solution: We observe that $\mathbb{E}(M_n|\mathcal{F}_{n-1}) = \mathbb{E}(X_n|\mathcal{F}_{n-1}) - n! = n \cdot X_{n-1} - n! = n(X_{n-1} - (n-1)!) = n \cdot M_{n-1} \neq M_{n-1}$.

∎

Problem 12.13 Find a stochastic differential for the process $X_t = (\sqrt{123} + \frac{1}{2}W_t)^2$, where (W_t) is a Wiener process.

Solution: Here $X_t = f(W_t)$ with the function $f(x) = (\sqrt{123} + \frac{1}{2}x)^2$. The first and second derivatives of this function are $f'(x) = 2(\sqrt{123} + \frac{1}{2}x)$, $f''(x) = 1$. Therefore, using the Ito formula, the stochastic differential of (X_t) is

$$dX_t = f'(W_t)dW_t + \frac{1}{2}f''(W_t)dt$$
$$= 2(\sqrt{123} + \frac{1}{2}W_t)dW_t + \frac{1}{2}dt.$$

∎

Problem 12.14 Let $(N_t)_{t \geq 0}$ be a Poisson process with intensity $\lambda = 1$. Prove that $(N_t - t)^2 - t$ is a martingale with respect to natural filtration generated by (N_t). Calculate $\mathbb{E} \int\limits_1^3 N_t dt \cdot \int\limits_2^4 N_t dt$

Hint: In the first case, please, check a martingale property. In the second case the answer is $34\frac{1}{3}$.

∎

Problem 12.15 Check whether the processes are martingales
(a) $X_t = W_t^3 - 3tW_t$;
(b) $X_t = W_t + 287t$;
(c) $X_t = e^{\frac{t}{2}} \sin(W_t)$,
where $(W_t)_{t \geq 0}$ is a Wiener process.

Solution:
(a) We have here $X_t = f(t, W_t)$ with the function $f(t, x) = x^3 - 3tx$. This function has the partial derivatives

$$\frac{\partial}{\partial x} f(t, x) = 3x^2 - 3t, \quad \frac{\partial^2 f(t, x)}{\partial x^2} = 6x,$$

$$\frac{\partial}{\partial t} f(t, x) = -3x.$$

Therefore, using the Ito formula, we derive

$$dX_t = (3W_t^2 - 3t)dW_t + \left(-3W_t + \frac{1}{2}(6W_t)\right) dt = (3W_t^2 - 3t)dW_t.$$

Therefore, (X_t) is a martingale as a stochastic integral has a martingale property.

In case (b) we have $\mathbb{E}X_0 = 0$ and for instance $\mathbb{E}X_1 = 287$. This implies that (X_t) cannot be a martingale, since martingales have constant expectations.

For (c) we have $X_t = f(t, W_t)$ with function $f(t, x) = e^{\frac{t}{2}} \cdot \sin(x)$ and its partial derivatives

$$\frac{\partial}{\partial x} f(t, x) = e^{\frac{t}{2}} \cdot \cos(x), \quad \frac{\partial^2 f(t, x)}{\partial x^2} = -e^{\frac{t}{2}} \cdot \sin(x),$$

$$\frac{\partial}{\partial t} f(t, x) = \frac{1}{2} e^{\frac{t}{2}} \cdot \sin(x).$$

Using the Ito formula we get

$$dX_t = e^{\frac{t}{2}} \cdot \cos(W_t)dW_t + \left[\frac{1}{2} e^{\frac{t}{2}} \cdot \sin(W_t) + \frac{1}{2}(-e^{\frac{t}{2}} \cdot \sin(W_t))\right] dt =$$

$$= e^{\frac{t}{2}} \cdot \cos(W_t)dW_t$$

which certifies that (X_t) is a martingale.

∎

Problem 12.16 Provide a condition on the mapping $\varphi : \mathbb{R}^1 \longrightarrow \mathbb{R}^1$ under which $\varphi(\tau)$ remains a stopping time, where τ is a stopping time.

Solution: Suppose the mapping φ meets the following conditions:
(a) φ is injective,

(b) $\varphi([0, \infty)) = [0, \infty)$,
(c) φ order-preserving and $t \leq \varphi(t)$ for all $t \in [0, \infty)$.
Condition (a) ensures that the inverse mapping $\varphi^{-1} : \varphi(\mathbb{R}^1) \longrightarrow \mathbb{R}^1$ is defined. Condition (b) ensures that $\varphi^{-1}(t)$ is well-defined and positive for all $t \geq 0$. Condition (c) together with the previous two conditions (a) and (b) gives that whenever τ is a stopping time
$$\{\varphi(\tau) \leq t\} = \{\tau \leq \varphi^{-1}(t)\} \in \mathcal{F}_{\varphi^{-1}(t)} \subseteq \mathcal{F}_t.$$

For instance, if $\varphi : [0, \infty) \longrightarrow [0, \infty)$ is a strictly increasing function satisfying $t \leq \varphi(t)$ for all $t \geq 0$, and φ is differentiable with $\varphi'(t) \geq 1$ for all t, $\varphi(0) = 0$, then by the mean value theorem we get $\varphi(t) \geq \varphi'(c) \cdot t \geq t$, and the above holds.

∎

Problem 12.17 Consider the function $p(t, x, y) = \frac{1}{\sqrt{2\pi t}} \exp\left\{-\frac{(x-y)^2}{2t}\right\}$, $x, y \in \mathbb{R}^1, t \in \mathbb{R}_t^1$, representing the transition density of a Wiener process (W_t). Prove the $p(t, x, y)$ satisfies to the PDE:
$$\frac{\partial}{\partial t} p(t, x, y) = \frac{1}{2} \frac{\partial^2}{\partial y^2} p(t, x, y).$$

Solution: On one hand side we have
$$\frac{\partial}{\partial t} p(t, x, y) = p(t, x, y) \left[-\frac{1}{2t} + \frac{(x-y)^2}{2t}\right].$$

On the other hand, differentiating with respect to y, we get
$$\frac{\partial^2}{\partial y^2} p(t, x, y) = p \cdot \frac{(x-y)^2}{t} + p \cdot \left(-\frac{1}{t}\right)$$
$$= p(t, x, y) \left[\frac{(x-y)^2}{t} - \frac{1}{t}\right].$$

Then it is clear that $p(t, x, y)$ satisfies the above differential equation.

∎

Problem 12.18 Prove that every non-negative local martingale is a supermartingale.

Hint: Apply the Fatou lemma.

∎

Problem 12.19 Let $(X_n)_{n=0,1,\ldots,N}$ be a submartingale with $E(X_N^+)^p < \infty$, $p > 1$. Prove the Doob inequality
$$E(\max_n X_n^+)^p \leq \left(\frac{p}{p-1}\right)^p E(X_N^+)^p, \quad X_N^+ = \max(0, X_N).$$

Solution: Denote $X_N^* = \max_{n \leq N} X_n^+$ and note that
$$\lambda \cdot P(X_N^* > \lambda) \leq EX_N^+ \cdot I_{\{X_N^* > \lambda\}}, \quad \lambda > 0.$$

Multiplying the above inequality by λ^{p-2} and integrating over $(0, \infty)$, we obtain

$$\int_0^\infty \lambda^{p-1}\mathbf{P}(X_N^* > \lambda)d\lambda \leq \mathbf{E}X_N^+ \cdot \int_0^{X_N^*} \lambda^{p-2}d\lambda \leq \frac{1}{p-1}\mathbf{E}X_N^+ \cdot (X_N^*)^{p-1}$$

Due to $\int_0^\infty \lambda^{p-1}\mathbf{P}(X_N^* > \lambda)d\lambda = \frac{1}{p}\int_0^\infty \lambda^p d\mathbf{P}(X_N^* \leq \lambda)$ we get $\mathbf{E}(X_N^*)^p \leq \frac{p}{1-p}\mathbf{E}X_N^+ \cdot (X_N^*)^{p-1}$, and with the help of the Hölder-inequality we derive

$$\mathbf{E}X_N^+ \cdot (X_N^*)^{p-1} \leq (\mathbf{E}(X_N^+)^p)^{\frac{1}{p}}(\mathbf{E}(X_N^*)^p)^{\frac{p-1}{p}}$$

and hence

$$\mathbf{E}(X_N^*)^p \leq \frac{p}{p-1}(\mathbf{E}(X_N^+)^p)^{\frac{1}{p}}(\mathbf{E}(X_N^*)^p)^{\frac{p-1}{p}}.$$

It leads to the desirable inequality.

∎

Problem 12.20 Give an example of a non-right-continuous filtration and a martingale which is not right-continuous.

Solution: For a positive real number $a > 0$ define $\Omega = \{a, -a\}$, $\mathbf{P}(a) = \mathbf{P}(-a) = \frac{1}{2}$, and the Bernoulli random variable

$$X = \begin{cases} a & \text{with probability } \frac{1}{2}, \\ -a & \text{with probability } \frac{1}{2}. \end{cases}$$

Define $X_t = \begin{cases} 0 & \text{if } t \leq t_0, \\ X & \text{if } t > t_0 \end{cases}$, $t \geq 0$, and $F_t = \begin{cases} \{\emptyset, \Omega\} & \text{if } t \leq t_0, \\ F_X = \sigma(X) & \text{if } t > t_0 \end{cases}$.

Process (X_t) is a martingale w.r. to (F_t). Both (X_t) and (F_t) are not right-continious.

∎

Problem 12.21 Let $(A_n)_{n\geq 0}$ be a predictable $(d \times d)$-matrix-valued sequence of random variables such that $\Delta A_n = A_n - A_{n-1}$ is positive-defined, $n \geq 1$, $\lambda\min(A)$ and $\lambda\max(A)$ are minimal and maximal eigenvalues of A. Let $(M_n)_{n\geq 0}$ be a d-dimensional square-martingale with the quadratic characteristic $<< M, M >>_n = (\langle M^i, M^j \rangle_n)_{i,j=1,...,d}$. Prove an analog of lemma 7.1 for d-dimensional case: (a.s)

$$\{\omega : \lambda_{\min}(A_\infty) = \infty\} \cap \{\omega : N \to\} \cap \{\omega : \limsup_n \frac{\lambda\max}{\lambda\min}(A_n) < \infty\} \subseteq \{\omega : A_n^{-1}M_n \to 0\}.$$

Hint: Adapt the proof of lemma 7.1 to this multidimensional case.

∎

Problem 12.22 Let (A_n) and (M_n) be as in the previous problem, $\lambda\min(A_n) \to \infty$ (a.s.), $\lim_n \sup \frac{\lambda\max}{\lambda\min}(A_n) < \infty$ (a.s.), and (a.s.)

$$\sum_{n=1}^\infty tr\left((A_n^{-1})\Delta << M, M >>_n (A_n^{-1})\right) < \infty.$$

12 Supplementary problems

Then (a.s.) $A_n^{-1} M_n \to 0, n \to \infty$.

Hint: Adapt the proof of theorem 7.1 to this multidimensional case.

■

Problem 12.23 Let us consider a polynomial transformation $P_n(t).Q_m(w_t)$ of a Wiener process $(w_t)_{t \geq 0}$, where P_n and Q_m are polynomials of degrees n and m relatively. Determine when this transformation leads to a martingale.

Hint: Use the Ito formula.

■

Problem 12.24 Let $(X_n)_{n=1,2,...}$ be a sequence of independent random variables with the density
$$f_a(x) = \begin{cases} e^{-ax} & if \ x \geq 0, \\ 0 & if \ x < 0. \end{cases}$$

Find the parameter a to provide that $Z_n = \prod_{i=1}^n X_i$ is a martingale w.r. to a natural filtration $(F_n^X)_{n=1,2,...}$.

Hint: Use the conditions $\int_0^\infty f_a(x)dx = 1$ and $\mathbf{E} X_n = 1$.

■

Problem 12.25 Assume $X_n \sim N(\mu_n, \sigma_n^2), n = 1, 2, ...,$ and $X_n \xrightarrow{L_2} X, n \to \infty$. Then $X \sim N(\mu, \sigma^2)$, where $\mu = \lim_{n\to\infty} \mu_n$, $\sigma^2 = \lim_{n\to\infty} \sigma_n^2$.

Solution: It follows from the L_2-convergence that $\mu_n \to \mu = \mathbf{E} X$ and $Var X_n \to Var X = \sigma^2, n \to \infty$. Hence, for an arbitrary $\lambda \in \mathbb{R}^1$ we obtain
$$\mathbf{E} e^{i\lambda X} = \lim_{n\to\infty} \mathbf{E} e^{i\lambda X_n} = \lim_{n\to\infty} e^{i\mu_n \lambda - \frac{\sigma_n^2}{2}\lambda^2} = e^{i\mu\lambda - \frac{\sigma^2}{2}\lambda^2}.$$

■

Problem 12.26 Let $b = b(x)$ and $\sigma = \sigma(x)$ be bounded functions from $\mathbb{R}^1$ to $\mathbb{R}^1$ such that $b \in \mathbb{C}^1(\mathbb{R}^1), \sigma \in \mathbb{C}^2(\mathbb{R}^1)$. Assume $(W_t)_{t \geq 0}$ is a wiener process. Then the Ito process
$$dX_t = b(X_t)dt + \sigma(X_t).dW_t,$$
can be rewritten in the form of the Stratonovich stochastic integral
$$dX_t = (b(X_t) - \frac{1}{2}\sigma(X_t)\sigma'(X_t))dt + \sigma(X_t).dW_t.$$

Show also that for $X_t = W_t$ and $F \in \mathbb{C}^2(\mathbb{R}^1)$ the Ito formula $F(X_t) - F(X_0) = \int_0^t F'(X_s)dX_s + \frac{1}{2}\int_0^t F''(X_s)ds$ admits the following form in terms of Stratonovich integral $F(X_t) - F(X_0) = \int_0^t F'(X_s).dX_s$.

Hint: Use definitions of these integrals.

■

Problem 12.27 Let us consider the exponential transformation of a probability measure $\mathbf{P}_T$, $T > 0$ to a measure function $\mathbf{P}_T^*$ such that

$$\frac{d\mathbf{P}_T^*}{d\mathbf{P}_T} = e^{aW_T + bT + c} = Z_T,$$

where $(W_t)_{t \leq T}$ is a Wiener process. Determine parameters a, b, and c under which $\mathbf{P}_T^*$ will be a probability measure.

Hint: Use the condition $\mathbf{E}Z_T = 1$.

∎

Problem 12.28 Prove that the first and second variations of a Wiener process $(W_t)_{t \geq 0}$ converge (a.s.) to infinity and to the length of time interval correspondingly.

Solution: For a fixed $T > 0$ we define a subdivision $t_i^n = iT \cdot 2^{-n}$, $i \leq 2^n$, $n \geq 1$ of the interval $[0, T]$. Define

$$FV^n = \sum_{i=0}^{2^n-1} |W_{t_{i+1}^n} - W_{t_i^n}| = \sum_{i=0}^{2^n-1} |\Delta W_{t_i^n}|$$

and

$$SV^n = \sum_{i=0}^{2^n-1} (\Delta W_{t_i^n})^2, \quad n \geq 1.$$

For SV^n using properties of (W_t) we have

$$\mathbf{E}(SV^n - T)^2 = \mathbf{E} \sum_{i=0}^{2^n-1} \left((\Delta W_{t_i}^n)^2 - T \cdot 2^{-n}\right)^2$$

$$= \sum_{i=0}^{2^n-1} \mathbf{E}\left((\Delta W_{t_i}^n)^2 - T \cdot 2^{-n}\right)^2$$

$$= 2^n \cdot 2 \cdot (T \cdot 2^{-n})^2$$

$$= 2 \cdot T^2 \cdot 2^{-n}$$

and

$$\left\| \sum_{n=1}^{\infty} |SV^n - T| \right\|_{L_2} \leq \sum_{n=1}^{\infty} \|SV^n - T\|_{L_2}$$

$$\leq \sqrt{2}T \cdot \sum_{n=1}^{\infty} 2^{\frac{-n}{2}} < \infty.$$

Hence, $SV^n \to T$ (a.s.) as $n \to \infty$ due to $\sum_{n=1}^{\infty} |SV^n - T| < \infty$ (a.s.).
Regarding FV^n for each $\omega \in \Omega$ we have $SV^n(\omega) \leq (\max_i |\Delta W_{t_i^n}(\omega)|)$.
$\sum_{i=0}^{2^n-1} |\Delta W_{t_i^n}(\omega)| \leq (\max_i |\Delta W_{t_i^n}(\omega)|) \cdot FV^n(\omega)$. It follows from this inequality for paths of $(W_t)_{t \geq 0}$ that $\lim_{n \to \infty} \inf FV^n(\omega)$ cannot be finite, otherwise $SV^n \xrightarrow[n \to \infty]{} 0$.

12 Supplementary problems

Problem 12.29 Let $(M_t)_{t \geq 0}$ be a continuous square integrable martingale with a finite first variation $FV_t(M)$. Show that $M_t = M_0$ (a.s.) for all $t \geq 0$.

Hint: Let (t_j^n) be a finite partition of $[0, t]$ with $\max_j \Delta t_j^n \to 0$ as $n \to \infty$. Then (a.s.)

$$\sum_j |\Delta M_{t_j^n}|^2 \leq FV_t(M(\omega)) \cdot \max_j |\Delta M_{t_j^n}(\omega)| \to 0, \quad n \to \infty,$$

and
$$<M, M>_t = 0 \quad (a.s.).$$

Hence, $M_t = M_0$ (a.s.).

Problem 12.30 Let $M = (M_t)_{t \geq 0}$ be a continuous local martingale and

$$Z_t(M) = \exp\{M_t - \frac{1}{2}\langle M, M \rangle_t\}, \quad t \in [0, \infty].$$

Assume that (Krylov's condition):

$$\liminf_{\varepsilon \to 0} \varepsilon \cdot \log \mathbf{E} \exp\{\frac{1-\varepsilon}{2}\langle M, M \rangle_\infty\} < \infty.$$

Prove that $\mathbf{E} Z_\infty(M) = 1$, i.e. the exponential local martingale $(Z_t(M))_{t \geq 0}$ is uniformly integrable. Show that the condition above is wider than the Novikov condition $\mathbf{E} e^{\frac{1}{2}\langle M, M \rangle_\infty} < \infty$.

Hint: First of all we note that $\mathbf{E} Z_\infty(M) \leq 1$ as a consequence of a supermartingale property of $(Z_t(M))$ and the Fatou lemma. Using the Hölder inequality we can derive for a constant $c > 0$ that

$$1 = \mathbf{E} Z_\infty((1-\varepsilon)M) = \mathbf{E} e^{(1-\varepsilon)(M_\infty - \frac{1}{2}\langle M,M \rangle_\infty)} \cdot e^{\frac{(1-\varepsilon)\varepsilon}{2}\langle M,M \rangle_\infty} \cdot I_{\{\langle M,M \rangle_\infty \leq c\}}$$
$$+ \mathbf{E} e^{(1-\varepsilon)(M_\infty - \frac{1}{2}\langle M,M \rangle_\infty)} \cdot e^{\frac{(1-\varepsilon)\varepsilon}{2}\langle M,M \rangle_\infty} \cdot I_{\{\langle M,M \rangle_\infty > c\}}$$
$$\leq (\mathbf{E} Z_\infty(M))^{1-\varepsilon} \cdot \left(\mathbf{E} e^{\frac{(1-\varepsilon)\varepsilon}{2}\langle M,M \rangle_\infty} \cdot I_{\{\langle M,M \rangle_\infty \leq c\}}\right)^\varepsilon$$
$$+ (\mathbf{E} Z_\infty(M) \cdot I_{\{\langle M,M \rangle_\infty > c\}})^{1-\varepsilon} \cdot \left(\mathbf{E} e^{\frac{(1-\varepsilon)\varepsilon}{2}\langle M,M \rangle_\infty}\right)^\varepsilon.$$

By taking the limit when $\varepsilon \to 0$ in the above inequality, we arrive to

$$1 \leq \mathbf{E} Z_\infty(M) + const \cdot \mathbf{E} Z_\infty(M) \cdot I_{\{\langle M,M \rangle_\infty > c\}}.$$

Taking the limit as $c \to \infty$, we get $\mathbf{E} Z_\infty(M) \geq 1$.

Problem 12.31 Show that every process (A_t) with finite variation can be represented as the difference of two increasing processes.

Hint: Use the representation

$$A_t = \frac{1}{2}(|A|_t + A_t - \frac{1}{2}(|A|_t - A_t)),$$

where $|A|_t$ is the variation of A on $[0, t]$.

∎

Problem 12.32 Let $(X_t)_{t \geq 0}$ be a semimartingale and let (A_t) be a process of finite variation. Prove that

$$[X, A]_t = \sum_{s \leq t} (\Delta X_t)(\Delta A_s).$$

In particular, $[X, A] = 0$, if (A_t) or (X_t) is continuous.

Hint: Use the limiting arguments dividing $[0, t]$ by a subdivision $t_j^n = jt2^{-n}$, $j = 0, ..., 2^n$, $n \geq 1$.

∎

Problem 12.33 Prove the Levy characterization of a Wiener process (Remark 8.1).

Solution: Let us prove that the following three statements are equivalent for a continuous martingale $(X_t)_{t \geq 0}$, $X_0 = 0$:
(1) Process (X_t) is a standard Wiener process on the underlying stochastic basis;
(2) Process $(X_t^2 - t)_{t \geq 0}$ is a martingale;
(3) Process $(X_t)_{t \geq 0}$ has a quadratic variation $[X, X]_t = t$.

To prove (1) $\Rightarrow$ (2) we just observe that $\mathbf{E}|X_t^2 - t| < \infty$ for each $t \geq 0$ and for $s \leq t$, $\mathbf{E}(X_t^2 - t|F_s) = \mathbf{E}((X_t^2 - X_s^2) + X_s^2 - (t - s) - s|F_s) = X_s^2 - s$ due to the properties of a Wiener process (X_t).

For the proof of the second implication (2) $\Rightarrow$ (3) we note that $(X_t^2)_{t \geq 0}$ is a submartingale with the Doob-Meyer decomposition $X_t^2 = (X_t^2 - t) + t$. Hence, $[X, X]_t = <X, X>_t = t$.

The third implication (3) $\Rightarrow$ (1) can be stated as follows. First, we prove that increments of $(X_t)_{t \geq 0}$ are Gaussian. Applying the Ito formula to $e^{i\lambda x + \frac{\lambda^2}{2}t} = f(t, x)$, $t \geq 0$, and $\lambda > 0$, we obtain

$$df(t, X_t) = \frac{\partial}{\partial t} f(t, X_t) dt + \frac{\partial}{\partial x} f(t, X_t) dX_t + \frac{1}{2} \frac{\partial^2}{\partial x^2} f(t, X_t) d[X, X]_t$$

$$= \frac{1}{2} \lambda^2 f(t, X_t) dt + i\lambda f(t, X_t) dX_t + (-\frac{\lambda^2}{2}) f(t, X_t) dt$$

$$= i\lambda f(t, X_t) dX_t,$$

and therefore, $(f(t, X_t))_{t \geq 0}$ is a martingale. Further

$$\mathbf{E}(e^{i\lambda X_t + \frac{1}{2}\lambda^2 t}|F_s) = e^{i\lambda X_s + \frac{1}{2}\lambda^2 s},$$

and

$$\mathbf{E} e^{i\lambda(X_t - X_s)} = \mathbf{E}(\mathbf{E}(e^{i\lambda(X_t - X_s)}|F_s)) = e^{-\frac{1}{2}\lambda^2(t-s)},$$

which means that $X_t - X_s \sim N(0, t - s), t \geq s$.
To finish the proof we show that the increments $X_{t_2} - X_{t_1}, ..., X_{t_n} - X_{t_n-1}$ are independent for any subdivision of time $0 \leq t_1 < t_2 < t_3 < ... < t_{n-1} < t_n$. We have

$$\mathbf{E}(e^{i\lambda_1 X_{t_1} + i\lambda_2(X_{t_2} - X_{t_1}) + ... + i\lambda_n(X_{t_n} - X_{t_n-1})}) = e^{-\frac{\lambda_1^2}{2}t_1} \times ... \times e^{-\frac{1}{2}\lambda_n^2(t_n - t_{n-1})},$$

which states independence of increments of (X_t). So, $(X_t)_{t \geq 0}$ is a Wiener process.

∎

Problem 12.34 Prove that any submartingale $X = (X_t)_{t \geq 0}$ on a standard stochastic basis $(\Omega, F, (F_t)_{t \geq 0}, P)$ admits the right-continuous modification, if EX_t is right-continuous.

Hint: Using limiting arguments together with a submartingale property we have $X_t \leq E(X_{t+}|F_t)$ (a.s.), $t \leq 0$. But $F_t = F_{t+}$, hence $X_t \leq X_{t+}$ (a.s.), and due to $EX_t = EX_{t+}$ we get the result.

∎

Problem 12.35 Let $M = (M_t)_{t \geq 0}$ be a martingale on a stochastic basis $(\Omega, F, (F_t)_{t \geq 0}, P)$. Show that a predictability property for the process $\phi = (\phi_t)$ is vital in getting the martingale property for a stochastic integral $\int_0^t \phi_s dM_s$.

Hint: Take a rich enough probability space (Ω, F, P) to accommodate two random variables: $\tau \geq 0$ with $P(\tau \leq t) = t \wedge 1$ and a Bernoulli random variable Y such that $P(Y = 1) = P(Y = -1) = \frac{1}{2}$. Define $M_t = Y.I_{\{\tau \leq t\}}$ and filtration $F_t = F_t^M$. In this case $\int_0^t M_s dM_s$ is not a martingale.

∎

Problem 12.36 Let continuous processes $A \in \mathcal{A}_{loc}^+$ and $M \in \mathcal{M}_{loc}$ are defined on a standard stochastic basis $(\Omega, F, (F_t)_{t \geq 0}, P)$. Assume that $b^i = b^i(x), x \in \mathbb{R}^1, i = 1, 2$, are bounded continuous functions and $X^i, i = 1, 2$, are (strong) continuous solutions of the stochastic differential equations w.r. to a semimartingale $Y_t = A_t + M_t, Y_0 = 0$:

$$dX_t^i = b^i(X_t^i)dA_t + dM_t,$$

where $X_0^i = x \in \mathbb{R}^1$.
Prove that the inequality $b^1(x) < b^2(x)$ for all $x \in \mathbb{R}^1$ implies $X_t^1(\omega) \leq X_t^2(\omega)$ (a.s.) for all $t \geq 0$.

Hint: Apply the method of proof similar to Lemma 9.1.

∎

Problem 12.37 Consider a stochastic differential equation as in problem 12.36

$$dX_t = b(X_t)dA_t + dM_t,$$

where $X_0 = x \in \mathbb{R}^1, t \leq T$.
Assume that $b^1 = b(x), x \in \mathbb{R}^1$ satisfies conditions of Theorem 9.3 and $A \ll \langle M, M \rangle$. Prove that this equation admits at least one (strong) solution.

Hint: Adapt the method of proof of Theorem 9.3 for this case.
∎

Problem 12.38 Let $dX_t = \mu dt + \sigma dW_t + v dN_t$, where W and N be a Wiener process and a Poisson process, respectively, with intensity $\lambda > 0$, $\mu, \sigma, v \in \mathbb{R}^1$. Let (τ_i) be the moments of the jumps of N. Prove the Ito formula for $F \in \mathbb{C}^2$:

$$F(X_t) = F(X_0) + \int_0^t F'(X_s)dX_s + \frac{1}{2}\int_0^t F''(X_{s-})\sigma^2 ds - \sum_{i=1}^{N_t} F'(X_{\tau_i-})\Delta X_{\tau_i} + \sum_{i=1}^{N_t}(F(X_{\tau_i}) - F(X_{\tau_i-})).$$

Hint: Adapt the Ito formula for semimartingales in this case.
∎

Problem 12.39 Let $N = (N_t)_{t \geq 0}$ be a Poisson process with intensity $\lambda > 0$ and $\alpha = (\alpha_t)$ be a bounded deterministic function.
Define the process

$$L_t = \exp\{\int_0^t \alpha_s d(N_s - \lambda s) + \int_0^t (1 + \alpha_s - e^{\alpha_s})\lambda ds\}$$

and prove that $L = (L_t^0)_{t \geq 0}$ is a martingale satisfying the equation

$$dL_t = L_{t-}(e^{\alpha_t} - 1)d(N_t - \lambda t).$$

Hint: Use the Ito formula.
∎

References

1. Baldi P.: An introduction through theory and exercises.tochastic calculus. universitext, (2017)
2. Beiglboeck M., Schachermayer W., and Veliyev B.: A short proof of the doob-meyer theorem. Stochastic Processes and their applications, vol. 122, no. 4, pp. 1204-1209, (2012)
3. Bishwal., Jaya PN.: Parameter Estimation in Stochastic Differential Equations. Springer-Verlag, Berlin- Heidelberg, (2008)
4. Borkar., Vivek S.: Stochastic approximation: A dynamical systems viewpoint. Cambridge University Press, Cambridge, (2008)
5. Borodin., A. N.: Stochastic processes. Birkhauser, (2018)
6. Bulinski, A. V., Shiryayev, A. N.: Theory of stochastic processes. Fizmatlit, Moscow, (2005)
7. Çinlar, E.: Probability and Stochastics. Springer, vol. 261, (2011)
8. Cohen, S. N., Elliott, R. J.: Stochastic calculus and applications. 2nd Edition, Springer-Science, (2015)
9. Doléans–Dade, C.: Stochastic processes and stochastic differential equations. in Stochastic Differential Equations, Springer-Verlag, pp. 7-73, (2010)
10. Durrett, R.: Essentials of Stochastic Processes. 3rd Edition. Springer, (2018)
11. Eberlein, E., and Kallsen, J.: Mathematical Finance. Springer, (2019)
12. Edgar, G. A., Sucheston, L.: Amarts: A class of asymptotic martingales. a. discrete parameter. Journal of Multivariate Analysis, vol. 6, pp. 193-221, (1976)
13. Etheridge, A.: A Course in Financial Calculus. Cambridge University Press, (2002)
14. Ikeda, N., and Watanabe, S.: Stochastic Differential Equations and Diffusion Processes. 2nd Edition. North-Holland, (1989)
15. Jacod, J., and Protter, P.: Probability Essentials. 2nd Edition. Springer, (2003)
16. Kallianpur, G., and Karandikar, R. L.: Introduction to option pricing theory. Springer Science & Business Media, (2012)
17. Karatzas, I., and Shreve, S.: Brownian motion and stochastic calculus. Springer, New York, (1998)
18. Klebaner, F. C.: Introduction to stochastic calculus with applications. World Scientific Publishing Company, (2012)
19. Kolmogorov, A. N.: Foundations of the Theory of Probability. 2nd Edition. Chelsea, New York, (1956)
20. Kruglov, V. M.: Stochastic Processes. Academy, Moscow, (2013)
21. Krylov, N. V.: Introduction to the theory of random processes. Providence: American Mathematical Soc., (2002)
22. Krylov, N. V.: Controlled diffusion processes. Springer-Verlag, (1980)

23. Lamberton, D., and Lapeyre, B.: Introduction to Stochastic Calculus Applied to Finance. Chapman & Hall/CRC, (1996)
24. Le Gall, J.-F.: Brownian motion, martingales, and stochastic calculus. Springer, (2016)
25. Liptser, R. Sh., and Shiryaev, A. N.: Statistics of random processes. Springer, 2nd Ed, (2001)
26. Liptser, R. Sh, and Shiryaev, A. N.: Theory of martingales. Kluwer Academic Publishers, (1989)
27. Melnikov, A. V.: On solutions of stochastic equations with driving semimartingales. Proceeding of the third European young statisticians meeting, Catholic University, Leuven, pp. 120-124, (1983)
28. Melnikov, A. V.: On strong solutions of stochastic differential equations with nonsmooth coefficients. Theory Probab. Appl., vol. 24, no. 1, pp. 146-149, (1979)
29. Melnikov, A. V.: On the theory of stochastic equations in components of semimartingles. Sbornik Math, vol. 38, no. 3, pp. 381-394, (1981)
30. Melnikov, A. V.: Stochastic differential equations: Singularity of coefficients, regression models and stochastic approximation. Russian Math Surveys, vol. 51, no. 5, pp. 43-136, (1996)
31. Melnikov, A. V., and Novikov, A. A.: Sequential inferences with fixed accuracy for semimartingales. Theory of Probability & Its Applications, vol. 33, no. 3, pp. 446-459, (1989)
32. Melnikov, A. V., and Shiryayev, A. N.: Criteria for the absence of arbitrage in the financial market. Frontiers in pure and applied probability II: proceedings of the Fourth Russian-Finnish Symposium on Probability Theory and Mathematical Statistics, pp. 121-134, (1996)
33. Meyer, P.-A.: Probability and potential. Blaisdell Publ.Company, (1966)
34. Nevel'son, M. B., and Has'minskii, R. Z.: Stochastic approximation and recursive estimation. AMS, Providence, (1976)
35. Øksendal, B.: Stochastic differential equations. 5th Edition. Springer, (2000)
36. Protter, P. E.: Stochastic Integration and Differential Equations. 2nd Edition. Springer, (2005)
37. Revuz, D., and Yor, M.: Continuous Martingales and Brownian Motion. 2nd Edition. Springer-verlag, (1999)
38. Schachermayer, W., and Teichmann, J.: How close are the option pricing formulas of bachelier and black-merton-scholes?. Mathematical Finance, vol. 18, no. 1, pp. 155-170, (2008)
39. Shiryaev, A. N.: Essentials of Stochastic Finance. World Scientific, (1999)
40. Shiryaev, A. N.: Probability. 2nd Edition. Springer, (1996)
41. Skorokhod, A. V.: Lectures on the theory of stochastic processes. Utrecht: VSP, (1996)
42. Tikhonov, A. N., Vasileva, A. B., and Sveshnikov, A. G.: Differential Equations. Springer, (1985)
43. Valkeila, E., and Melnikov, A. V.: Martingale models of stochastic approximation and their convergence. Theory of Probability & Its Applications, vol. 44, no. 2, pp. 333-360, (2000)
44. Wentzell, A. D.: A Course in the Theory of Stochastic Processes. McGraw-Hill, (1981)
45. Williams, D.: Probability and Martingales. Cambridge University Press, (1991)

Index

A
Absolute continuity, 41
    local, 74
Absolute continuity of measures, 41
Accessible stopping time, 140
Algebra, 2
Atom, 46
Autoregression model, 68, 187

B
Bachelier
    discrete model, 75
    formula, 75
Bank account, 75
Bernoulli distribution, 7
Binomial market model, 180
Black-Scholes
    formula, 129
    model, 129
Borel function, 143
Borel space, 5, 13
Borel-Cantelli lemma, 89
Brownian motion, 82

C
Cadlag process, 142
Call option, 127, 128
Cantor function, 9
Capital of strategy, 176
Cauchy-Schwartz inequality, 27
Central Limit Theorem, 31
Change of measure, 53
Change of time, 50
Change of variables formula, 19

Characteristic function, 36
Chebyshev inequality, 26
Class D, 139
Comparison theorem, 113
Compensator, 52, 146, 147
Complete probability space, 83
Conditional expectation, 43
Controlled diffusion process, 132
Convergence
    weak, 38
Convergence of Random Variables, 28
Cox-Ross-Rubinstein model, 180
Cylinder, 6

D
Debut, 142
Diffusion coefficient, 120
Diffusion process, 120
Dirichlet function, 19
Discrete distribution, 7
Discrete stochastic integral, 53
Distribution
    Bernoulli, 7
    Binomial, 7
    density function, 10
    discrete, 7
    finite-dimensional, 118
    function, 7
    Normal, 8
    Poisson, 8
    Uniform, 7, 8
Doleans exponent, 168
Doob decomposition, 52
Doob inequalities, 144

Doob-Meyer decomposition, 146
Downcrossing, 58
Drift coefficient, 114

E
Elementary event, outcome, 1
Equivalence of measures, 104
Existence of a solution, 109
Expectation, 18
Extended Random Variable, 13

F
Fair price, 127
Fatou lemma, 22
Feynman-Kac representation, 123
Filtration, 50
Financial market, 75, 129
Finite-additive measure, 2
Finite-dimensional
    process, 97
Finite-dimensional distribution, 118
First variation, 86
Fisk decomposition, 158
Fixed accuracy property, 175
Fokker-Planck equation, 121
Formula
    Bachelier, 76
    Ito, 98
    Merton, 180
Function
    Borel, 15
    Cantor, 9
    characteristic, 36
    Dirichlet, 19
Functional space, 81
Fundamental sequence, 93

G
Gaussian process, 90
General theory of stochastic processes, 139
Generalized martingale, 80
Generator, 120
Geometric Brownian Motion, 112, 129
Girsanov theorem, 74
Graph, 140
Gronwall lemma, 109

H
Haar system, 88
Hamilton-Jacobi-Bellman principle, 133

Hamilton-Jacoby-Bellman equation, 138
Helly principle, 32

I
Incomplete market, 176
Independent Random Variables, 16
Indicator, 13
Indistinguishable processes, 141
Inequality
    Cauchy-Schwartz, 26
    Chebyshev, 26
    Jensen, 26
    Kolmogorov-Doob, 56
    Kunite-Watanabe, 160
Information flow, 50
Integrable, 18
Integral
    Lebesgue, 19
Interval of non-arbitrage prices, 135
Inverse image, 13
Isometric property, 92
Ito formula, 98
Ito process, 98
Ito stochastic integral, 97

J
Jensen inequality, 26
Joint Quadratic characteristic, 53
Jump-diffusion model, 180

K
Kolmogorov
    backward, forward equation, 121
    consistency theorem, 6
    variance condition, 67
Kolmogorov-Chapman equation, 118
Kolmogorov-Doob inequality, 56
Kunita-Watanabe
    decomposition, 54
    inequality, 160
Kurtosis, 20

L
Large Numbers Law, 28
Least-squares estimate, 68
Lebesgue
    dominated convergence theorem, 23
    integral, 19
Lebesgue measure, 5
Levy theorem, 64

Index

Lindeberg condition, 39
Lipschitz condition, 108
Local Lipschitz conditions, 108
Local martingale, 80, 151
Localization, 96
Localizing sequence, 80
Locally square-integrable martingale, 151
Lower option price, 135
LS-estimate, 68

M

Markov process, 117
Martingale, 51
    generalized, 80
    local, 80, 151
    locally square-integrable, 151
    Square integrable, 52
Martingale difference, 51
Martingale measure, 127
Martingale representation, 104
Mathematical finance, 112, 126, 127
Maturity time, 127
Measurable
    mapping, 16
Measurable space, 4
Measure, 1
    finite-additive, 2
    Lebesgue, 5
    Martingale, 127
    Wiener, 11
Merton formula, 180
Method of monotonic approximations, 113
Modification
    continuous, 84
    right-continuous, 87
Monotonic class, 4
Multiplication rule, 182

N

Normal distribution, 8
Novikov condition, 75

O

Optimal control, 133
Option
    call, 75
Optional Sampling Theorem, 55
Optional sigma-algebra, 142
Ornstein-Uhlenbeck process, 112

P

Parseval identity, 90
Partial Differential Equation, 123
Poisson distribution, 8
Polynomial, 101
Predictable process, 143
Predictable sequence, 52
Predictable sigma-algebra, 142
Predictable stopping time, 141
Principle of optimality, 133
Probability measure, 3
Probability space, 3, 5
Process with finite variation, 143
Progressively measurable, 98
Prokhorov theorem, 34
Purely discontinuous martingale, 150
Put-call parity, 129

Q

Quadratic bracket, 151
Quadratic characteristic, 52, 151
Quasimartingale, 79

R

Radon-Nikodym
    density, derivative, 42
    theorem, 42
Random set, 142
Random variable, 13
    extended, 13
    uniformly integrable, 23
Regression
    analysis, 67
    function, 71
Regression model, 68
Relative compactness, 34
Replicating strategy, 176

S

Sample path, 49
Scalar product, 88
Schauder system, 88
Semimartingale, 155
Sequential estimate, 132
Sigma-algebra, 4
    optional, 142
    predictable, 142
Singular distribution, measure, 8
Skewness, 20
Small perturbations method, 136
Step function, 91

Stochastic Approximation, 71, 183
Stochastic basis, 55
Stochastic calculus, 159
Stochastic Differential Equation, 107
Stochastic exponential, 54, 168
    multiplication rule, 168
Stochastic interval, 142
Stochastic Kronecker's Lemma, 66
Stochastic regression analysis, 182
Stochastic volatility, 134
Stochastically continuous process, 87
Stock price, 46, 112
Stopped process, 154
Stopping time, 50
    accessible, 140
    graph, 140
    predictable, 140
    totally inaccessible, 141
Strategy, 176
Stratonovich integral, 97
Strong solution, 113
Subdivision, 83
Submartingale, 52
Supermartingale, 52

T
Theorem
    Caratheodory, 5
    CLT, 38
    Comparison, 113
    Girsanov, 74

    Kolmogorov consistency, 6
    Lebesgue dominated convergence, 27
    Levy, 64
    Optional Sampling, 55
    Prokhorov, 34
    Radon-Nikodym, 42
Theory of Probability, 1
Tightness, 34
Totally inaccessible stopping time, 141
Transition probability function, 118

U
Uniform distribution, 8
Uniformly integrable random variable, 23
Uniqueness of solutions, 112, 132
Upcrossing, 57
Upper option price, 135
Usual conditions, 139, 144

V
Value function, 133
Variance, 20

W
Wald identity, 95
Weak convergence, 38
Weak solution, 113
Wiener measure, 11

The manufacturer's authorised representative in the EU is Springer Nature Customer Service Centre GmbH, Europaplatz 3, 69115 Heidelberg, Germany. If you have any concerns regarding our products, please contact ProductSafety@springernature.com

Printed and bound by CPI Group (UK) Ltd, Croydon, CR0 4YY
25/03/2026
02078169-0001